Trends of Environmental Forensics in Pakistan

Trends of Environmental Forensics in Pakistan

Edited by

Shazia Iftikhar, PhD
Associate Professor
Department of Environmental Sciences
Fatima Jinnah Women University
The Mall, Rawalpindi, Pakistan

ELSEVIER

ACADEMIC PRESS
An imprint of Elsevier

Academic Press is an imprint of Elsevier
125 London Wall, London EC2Y 5AS, United Kingdom
525 B Street, Suite 1650, San Diego, CA 92101, United States
50 Hampshire Street, 5th Floor, Cambridge, MA 02139, United States
The Boulevard, Langford Lane, Kidlington, Oxford OX5 1GB, United Kingdom

TRENDS OF ENVIRONMENTAL FORENSICS IN PAKISTAN

ISBN: 978-0-12-819436-2

Publisher: Stacy Massuci
Acquisition Editor: Elizabeth Brown
Editorial Project Manager: Emily Thomson
Production Project Manager: Kiruthika Govindaraju
Cover Designer: Alan Studholme

Contents

Contributors

Khuram Shahzad Ahmad, PhD
Department of Environmental Sciences, Fatima Jinnah Women University, Rawalpindi, Pakistan

Humaira Arif, MPhil
Department of Environmental Sciences, Fatima Jinnah Women University, Rawalpindi, Pakistan

Irum Asif, MPhil
Department of Environmental Sciences, Fatima Jinnah Women University, Rawalpindi, Pakistan

Marria Ghalib, MPhil
Department of Environmental Sciences, Fatima Jinnah Women University, Rawalpindi, Pakistan

Palwasha Gul, MPhil
Department of Environmental Sciences, Fatima Jinnah Women University, Rawalpindi, Pakistan

Rija Hafeez, MPhil
Department of Environmental Sciences, Fatima Jinnah Women University, Rawalpindi, Pakistan

Hareem, MPhil
Department of Environmental Sciences, Fatima Jinnah Women University, Rawalpindi, Pakistan

Shazia Iftikhar, PhD
Department of Environmental Sciences, Fatima Jinnah Women University, Rawalpindi, Pakistan

Shaan Bibi Jaffri, PhD Scholar
Department of Environmental Sciences, Fatima Jinnah Women University, Rawalpindi, Pakistan

Sara Majid, PhD Scholar
Department of Environmental Sciences, Fatima Jinnah Women University, Rawalpindi, Pakistan

Zoobia Nadeem, MPhil
Department of Environmental Sciences, Fatima Jinnah Women University, Rawalpindi, Pakistan

Zainab Naeem, MPhil
Department of Environmental Sciences, Fatima Jinnah Women University, Rawalpindi, Pakistan

Arooba Nosheen, MPhil Scholar
Department of Environmental Sciences, Fatima Jinnah Women University, Rawalpindi, Pakistan

Anam Qaisar, MPhil
Department of Environmental Sciences, Fatima Jinnah Women University, Rawalpindi, Pakistan

Shania Sharif, PhD Scholar
Department of Environmental Sciences, Fatima Jinnah Women University, Rawalpindi, Pakistan

Nageen Sohail, MPhil
Department of Environmental Sciences, Fatima Jinnah Women University, Rawalpindi, Pakistan

Preface

This project is the assemblage of the reality of environmental forensics in Pakistan. The idea for this project originates from the need for baseline data with regard to environmental crisis in Pakistan. Over the period of 10 years, I have been teaching environmental forensics to postgraduates; however, the state of environmental forensics in Pakistan is still undefined. The alterations in the soil chemistry is due to malicious effects of pesticides; the vicissitudes in biosphere and hydrosphere due to deviating toxicological responses; advent of biological warfare and its effects across the globe and exclusively in Pakistan are not in the limelight, therefore there is a strong need to highlight them at international forums.

This book is the first-ever project reporting the criminal events and their mitigation measures in Pakistan. The biggest hurdle faced while writing this book was access to authentic data, which was overcome over time. The evidence provided in this book is concrete and has been peer-reviewed. The language used in this book has been kept simple, effective for both scientific and nonscientific communities.

I am highly thankful to Elsevier publisher and complete editorial team that provided their consistent efforts and time for making the publication of this book possible. I am also highly thankful to my husband, Prof. Dr. Ifthikhar Ahmad, who has always remained a source of never ending inspiration and motivation for me.

Dr. Shazia Iftikhar
DAAD and Alexander von Humboldt Fellow
Associate Professor
Department of Environmental Sciences
Fatima Jinnah Women University
The Mall, Rawalpindi
Pakistan

CHAPTER

Environmental toxicology: fundamental and forensic

1

Rija Hafeez, MPhil, Zoobia Nadeem, MPhil, Shazia Iftikhar, PhD
Department of Environmental Sciences, Fatima Jinnah Women University, Rawalpindi, Pakistan

Chapter outline

Abstract

Toxicology plays an immense role in the field of environmental forensics, especially for the identification and assessment of hazardous chemicals such as drugs and heavy metals. However, to apply this study for environmental forensic investigations, the investigator must be equipped with prior knowledge regarding chemical toxicology. Therefore, this chapter aims to equip the reader with the fundamental knowledge for carrying out toxicological investigations by briefly describing some basic concepts of toxicology such as classification of toxicants,

Trends of Environmental Forensics in Pakistan. https://doi.org/10.1016/B978-0-12-819436-2.00001-7

sources and routes of exposure, dispositioning of toxic chemicals, dose-response relationship, and hazard and risk assessment. Furthermore, the chapter also summarizes some basic concepts regarding toxicological investigations such as types of samples and procedure for sample collection, preservation, storage, and analysis. Case studies from Pakistan have also been highlighted to provide the reader with practical application of analytical toxicology for environmental forensic.

Keywords: Analytical techniques; Analytical toxicology; Chemical toxicology; Environmental forensics; Environmental monitoring; Toxicological investigation.

Graphical Abstract

1.1 Introduction

Toxicology includes the study of the harmful connections between chemicals and biological systems. All living things are being exposed to massive variety of chemicals. These chemicals range from inorganic chemicals, metals to large multifarious organic molecules; however, they are all theoretically toxic. The study of pathological, biochemical, and physiological effects of harmful substances is the interesting brief of the toxicologist. Toxicology is a multidisciplinary subject which incorporates many areas as shown in Fig. 1.1. This makes it a captivating and stimulating area of research.

Today's challenge of toxicology is to transmit basic biochemical, pathological, chemical, and physiological knowledge along with experimental observation to advance an understanding of certain substances that cause the disturbance in a biological system which may lead to toxic effects.

Awareness to the delinquent of human and animal exposure to toxic chemicals, in our present environment, has grown in recent years. Rachel Carson with her book *Silent Spring* was one of the first to bring this issue to the attention of the public. This was a description of the devastating effects of pesticides on the environment of the North America, stipulating flora and fauna. Therefore, toxicology has another dimension: the social, ethical, and legal aspects of exposure of populations to chemicals of unknown or uncertain hazard. Hazard and risk assessments (RAs) and value judgments become important. The toxicologist is often asked to make such assessments and judgments. So, toxicology has a very important role to play in modern society, and consequently it is now growing rapidly as a new subject (Timbrell, 2008).

1.1.1 Types of toxic substances

Toxic substances are of several classes in relation to the mode a living organism is exposed to them such as drugs, pesticides, food additives, industrial chemicals, natural toxins, environmental pollutants, and household poisons.

Most of the people in the developed countries devour drugs throughout their lives and suffer from unembellished diseases and infections. Whereas

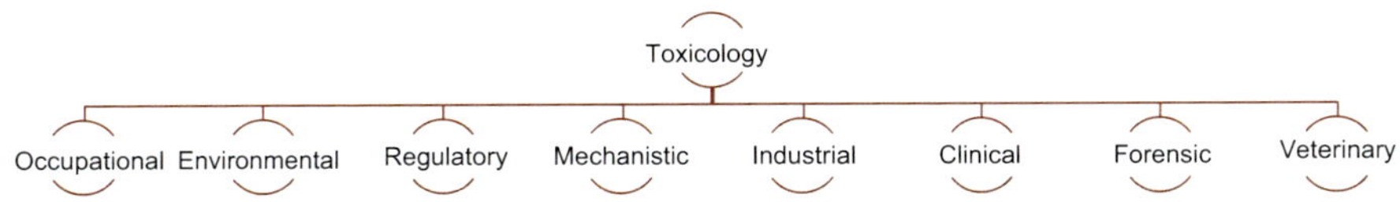

FIGURE 1.1 Schematic diagram of multidisciplinary branches of toxicology.

drugs have been designed to be highly potent in disturbing the biological systems, and consequently many are extremely toxic. Drug toxicity may result because of either an overdose or a rare and unusual adverse effect because of weak immune system. There are many drugs which are also beneficial for the health. Man-made chemicals are very complex, either they can be good or bad, depending upon their composition or a particular dose required for its usage. Drugs vary immensely in chemical structure and possess a wide variety of biological activities. Sometimes these drugs are considered as the foreign substances of known biological activity that a person ingests deliberately. Alcohol and cigarettes are both used intentionally most of the times because of their biological activity and both possess toxic properties (Frank and Ottoboni, 2011).

Different types of components in natural food are toxic to different individuals. Food additives are generally of low biological activity. Many different additives are now added to food to change the flavor of the food. These additives also change color, prevent spoilage, and sometimes change the nature of the food (Altug, 2002). These food additives can result into adverse reactions as mentioned below in Table 1.1.

Industrial chemicals do affect the environment in one way or another and are a great source of environmental pollution. They are dangerous within the workplace where they are utilized, formulated, and manufactured. In the comprehensive manner, industrial exposure includes exposure to the solvents used in photocopiers and typists' correction fluid. Though in general exposure is controlled by law, realistic levels may still prove to be hazardous in the long term and acute exposure due to accidents will always occur. The occurrence of diseases such as cancer often makes it difficult to regulate the cause until sufficient workforce have presented with the disease for the connotation with the toxic compound to be made (Timbrell, 2008).

The main sources of pollution in the environment are industrial processes that deliberately release pollutants such as pesticides, particulate matter, greenhouse gases. The most evident pollutant in the atmosphere is smoke from factories and power stations. Many plants and animals produce toxic substances for self-justifying and aggressive purposes. Natural toxins of animal, plant, and bacterial origin involve an extensive variety of chemical types and are a significant cause of human poisonings. Household poisons include substances such as drugs, pesticides, and solvents. These types of compounds are usually acute instead of chronic, when exposed. Many of the household items used for cleaning are irritants, and some are corrosive.

Table 1.1 Adverse reactions: causes and foods commonly involved.

Type of reaction	Cause	Food/food compounds involved
Food allergy	Allergic reaction to a food caused by an oversensitive reaction of the body's immune system	Eggs, milk, wheat, fish, shellfish, nuts, peanuts, soybean, rice
Pharmacological	Absorption of certain amines from foods containing high amounts Other substances with pharmacological type of action	Fermented foods (cheese, red wine, sauerkraut, fermented sausages), fish products Caffeine
Enzyme defects	Failure of normal enzymatic breakdown after absorption Failure to digest, so that unabsorbed substances reach the lower intestine, where they are fermented	Alcohol, fructose, amines Lactose (in milk), fats
Irritant	Often made worse by acid which refluxes from the top end of the stomach to cause heartburn	Strong spices and flavors, sulfites
Toxic	Toxins	Some shellfish, badly stored food (e.g., green potatoes), some fungi
Psychological food intolerance	Emotional reaction to a food (reaction does not occur if food is not recognized)	

1.1.2 Classification of exposure

A chemical's physical state and the route of exposure determines its toxicity, it also often controls the route of exposure. Exposure is usually defined as a measurement of the level at which one encounters any substance. That substance could be a glass of a fine wine or a medication that we ingest or other substances that one inhales, applies on the skin, or injects directly into bloodstream. Many people are concerned with exposures to chemical substances in food or water, or in the air that one inhales, or in the environment that the skin is exposed to.

There are four routes by which a material can enter the body: inhalation, skin absorption, ingestion, and injection.

1.1.2.1 Inhalation

For most chemicals, which are in the form of vapors, mists, gases, or particulates, inhalation is the major route of entrance. Once inhaled, chemicals are either exhaled or deposited in the respiratory tract. If deposited, damage can occur through direct contact with tissue or the chemical may prolix into the blood through the lung-blood lining. Upon contact with the tissue in the upper respiratory tract or lungs, chemicals can cause health effects ranging from simple irritation to severe tissue destruction. Substances absorbed into the blood are circulated and distributed to organs that have an empathy for that particular chemical. Nanoparticles are being established as carriers for several medicinal compounds. These particles range from 1 to 100 nm (nm), and many of them cannot be absorbed because of their small size. Still, nanoparticles can accumulate in the bronchi to form larger particles that can then be absorbed. Acute health effects can then occur in the organs, which are sensitive to the toxicants (Frank and Ottoboni, 2011).

1.1.2.2 Skin absorption

Skin or dermal contact can cause effects that are relatively harmless such as redness or mild dermatitis and can affect more severely such as obliteration of skin tissue or any other devastating conditions. Many chemicals can also cross the skin barrier and be absorbed into the blood system. Once absorbed, they may produce systemic damage to internal organs. The eyes are particularly sensitive to chemicals. Even a short exposure can cause severe effects to the eyes or the substance can be absorbed through the eyes and be ecstatic to other parts of the body causing harmful effects. As a general rule, the majority of inorganic chemicals are not absorbed through intact skin. Organic chemicals may or may not be absorbed, depending on a number of conditions. Usually, an organic chemical in dry powder form is absorbed to a lesser degree than the same chemical in an aqueous solution, suspension, or paste (Frank and Ottoboni, 2011).

1.1.2.3 Ingestion

The primary way that chemicals enter our bodies is by ingestion. Chemicals that unintentionally get into the mouth and are swallowed do not generally harm the gastrointestinal tract itself unless they are nauseating or corrosive. Chemicals that are insoluble in the fluids of the gastrointestinal tract (stomach, small and large intestines) are mostly excreted, whereas others that are soluble are absorbed through the lining of the gastrointestinal tract and then transported by the blood to internal organs where they can cause damage. Nitroglycerine, for example, can be absorbed through the mucous membranes of the mouth, hence its medicinal administration by placement under the tongue to relieve the pains of angina. Some chemicals, such as ethyl alcohol, are absorbed from the stomach, as well as the small intestines. That is why the effects of alcoholic beverages are felt so rapidly and why food in the stomach can help delay those effects (Frank and Ottoboni, 2011).

1.1.2.4 Injection

Substances may enter the body, if the skin is breached or wounded by contaminated objects. Effects can then occur as the matter is circulated in the blood and settled in the target organs.

In actual practice, exposure to a chemical may or may not involve just one route. Dermal exposure can also become ingestion exposure when hands are not washed after handling or working with chemicals and before eating or smoking. Inhalation exposure can also become ingestion exposure when some of the chemical deposited on the walls of the nasopharynx is swallowed along with the secretions from that cavity, or when it is coughed up from the lungs and swallowed. Thus, a specified route of exposure is primarily, rather than solely, by that route (Frank and Ottoboni, 2011).

1.1.3 Dose-response relationship

Evaluating clinical effects based on the amount of exposure is a basic toxicology principle called dose-response. The dose is the total amount of chemical absorbed during an exposure. Dose depends on the concentration of the chemical and duration (contact time) of the exposure. Chemicals cause predictable toxic effects based on the dose (Madsen et al., 2017).

Ethanol is a good example. Incremental increases in blood ethanol levels result in predictable increases in alteration of consciousness (signs of inebriation), poor coordination, and eventually coma/respiratory depression, and finally death.

Dose-response analysis is determining the relationship between dose and the probability or the occurrence of effect. The complexity of this step in many contexts derives mainly from the need to extrapolate results from experimental animals (e.g., mouse, rat) to humans, and/or from high to lower doses. In addition, the differences between individuals due to genetics or other factors mean that the hazard may be higher for particular groups called susceptible populations (Timbrell, 2008).

The amount of chemical entering the body is the dose. This is usually given as milligram of chemical/kilogram of body weight = mg/kg. The dose is dependent upon environmental concentration, properties of the toxicant, frequency of exposure, length of exposure, and exposure pathway.

1.1.4 Disposition of toxic substances

Absorption is defined as the process of movement of an unchanged compound from its site of administration or exposure to the bloodstream. Most hazardous substances must gain access to the systemic circulation to exert their toxic effects through interaction with one or more internal organs, except those that cause a local reaction at the site of exposure. During absorption, toxic substances gain entrance to the body from the external environment by crossing cellular barriers (Gupta, 2016).

The schematic representation of various steps involved during absorption, distribution, and excretion and the possible toxicokinetic fate of a chemical after exposure by inhalation, dermal contact, and ingestion are presented in Fig.1.2.

Distribution can be defined as a process by which toxic substances move throughout the body and reach their site of action that is extracellular fluid and tissues. Once absorbed, a toxicant usually enters the interstitial fluid at the place of absorption and then passes into the tissue cells or enters the blood and/or lymph.

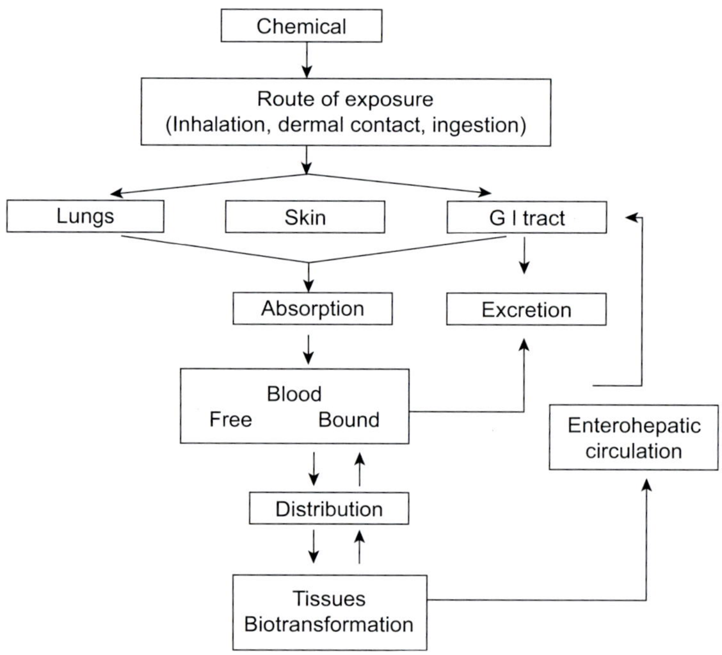

FIGURE 1.2 Schematic diagram of absorption, distribution, and excretion.

Factors that determine a compound's rate and extent of distribution therefore include the following:

- Molecular size (i.e., physicochemical properties of compound)
- Lipophilicity
- Plasma protein and tissue binding
- Blood flow and organ size
- Special compartmental and barriers (e.g., blood-brain barrier, blood-cerebrospinal barrier, placental barrier, and other barriers)
- Availability of special transport system
- The ability to interact with transmembrane transporter proteins
- Disease state

After absorption into the bloodstream, the chemicals penetrate into the various fluid compartments: (1) plasma, (2) interstitial fluid, (3) transcellular fluid, and (4) cellular fluid (Gupta, 2016).

Excretion is a process by which toxicants or their metabolites are permanently transferred from the body to the external environment. Therefore, excretion is one of the primary mechanisms of protecting the body from the toxic effects of toxicants through the eradication of these compounds from the body. Compounds that are rapidly eliminated are less likely to accumulate in tissues and damage critical cells. Although the terms elimination and excretion are sometimes used synonymously, the former term encompasses all the processes that decrease the amount of parent compound in the body, including biotransformation. Excretion by principal organs is called renal excretion; excretion by organs other than the kidneys is known as extrarenal or nonrenal excretion (Gupta, 2016).

1.1.5 Hazard and risk assessment

When it comes to chemical management, a distinction must be made between hazard and risk. Hazard defines the inherent property of a chemical having the potential to cause adverse effects when an organism, system, or population is exposed to that agent. Risk establishes the probability of the adverse effect occurring. We often perform toxicological research to understand the mechanisms and associated health risks following exposure to hazardous agents. The problem of understanding the hazards and risks associated with unintentional or coincidental exposures to chemicals is a complex one, with multiple uncertainties. Therefore, RA is a systematic scientific characterization of potential adverse health effects following exposure to these hazardous agents. The RA activities are designed to identify, describe, and measure qualities and quantities from these toxicological

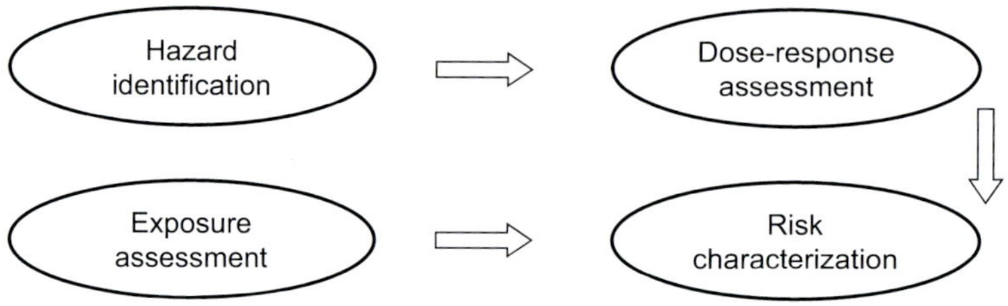

FIGURE 1.3 The four major elements of risk assessment.

studies, which are often conducted with homogeneous animal models at doses and exposure durations not encountered in a more heterogeneous human population (Gupta, 2016). There are four components of RA (Fig. 1.3).

In life-cycle assessment (LCA), it is necessary to compare quantities of chemicals released into the environment in terms of risks and values of toxicological effects. Many current methods trust directly on adaptations of monitoring-orientated approaches. The resultant hazard-based indicators reflect differences between the exposure in the region for an emission and differences in regulatory limits. Such regulatory hazard-based indicators, though, may not provide a reliable basis for relative assessment across chemicals in terms of toxicological danger, as they were not designed for this application purpose. It is therefore essential to consider other methods in LCA to provide comparative estimates, taking into account the full extent of toxicological risks and differences in consequences. This chapter provides a step-by-step description of the methodological similarities and differences between such risks and hazard-based indicators for LCA. An example for benzopyrene demonstrates a risk-based methodology, highlighting relationships with regulatory approaches and problems that remain in current practice (Pennington et al., 2006). Advances in toxicology are convinced to improve the quality of RAs and hazard identification to find substitute data for assumptions, and to help describe the models with more integrity (Gupta, 2016).

1.2 Analytical toxicology in environmental forensics

Analytical toxicology is the branch of science that is associated with the determination, identification, and measurement of drugs and other foreign substances (i.e., xenobiotics) and their metabolites in samples. This branch of science finds applications in a variety of sciences such as clinical

Table 1.2 Three steps involved in analytical toxicological investigation.

Pre-analytical	Record details regarding the poisoning episode and include circumstantial evidence as well as results obtained from investigation. Moreover, include the victim's medical and occupational history. Decide the priorities for analysis.
Analytical	Perform analysis
Post-analytical	Interpret the results and perform additional analysis, if required. Save any unused samples in case they are required for further tests.

toxicology, drug abuse screening, therapeutic drug monitoring, and occupational and environmental toxicology (Flanagan et al., 2008). Table 1.2 tabulates the steps that are usually involved in an analytical toxicological study.

In occupational and environmental toxicology, analytical toxicology can be used to monitor exposure to toxicants such as heavy metals (for example, lead [Pb] and mercury [Hg]), food-derived poisons (for example, botulinum toxins), drugs, and pesticides. Although the exposure to such chemicals is usually analyzed by monitoring of ambient environment, biological monitoring is also essential whenever effect monitoring cannot be performed. Biological monitoring involves the measurement of concentration of a compound and its metabolites in breath, blood, or urine. For instance, measurement of blood zinc protoporphyrin can provide an indication of Pb exposure. Moreover, in the case of chemical release incidents, toxicological analysis can not only be valuable in providing evidence regarding the exposure but also in allaying public apprehension where the exposure was insignificant (Flanagan et al., 2008).

1.2.1 Sample collection and handling

The collection and handling of samples must be carried out with the analysis technique in mind. Therefore, pr-hand knowledge is required regarding the nature of sample, nature of sample matrix, history of sample, and the circumstances in which analysis has to be carried out. While sampling, care must be taken regarding the changes that may take place in a sample due to human error or natural processes. Some of the main requirements for the sample include that it must represent the concentration of desired analyte in the whole sample matrix and should have appropriate size for analysis. For use in forensic investigations, sample integrity must be maintained to avoid the results from being questionable. Therefore, such samples must be protected during transport using tamper-evident seals and should be submitted to the laboratory by the investigating personnel (Flanagan et al., 2008). Tables 1.3 and 1.4 tabulate some of the biological samples and environmental toxicants along with their sampling procedure. For further details regarding sampling of environmental toxicants, consult *Interpol Pollution Control Forensic Investigation Manual*.

During sampling, health and safety measures also need to be assured as the samples may be hazardous in nature. For instance, biological samples may comprise infectious agents and should therefore be handled carefully by well-trained staff (Flanagan et al., 2008). Where hazardous samples are to be collected from environmental media, the sampler should be well-trained, should conform to the guidelines and protocol, and should wear protective gear such as protective glasses and gloves (INTERPOL, 2014).

1.2.2 Sample preparation and analysis

The complexity of the sample preparation procedure depends on a variety of factors such as nature of sample, concentration of analyte, and analytical technique employed for analysis. Some analytical techniques such as immunoassay methods may be performed directly, while others such as chromatographic methods require sample cleanup. In some cases, sample preparation may involve the removal of insoluble residues and interfering compounds and concentration or dilution of sample for increasing selectivity (Flanagan et al., 2008).

Sample preparation for solid samples may be carried out via physical or chemical disruption. Physical disruption may employ use of techniques such as sonication (use of ultrasonic waves), heating using microwaves, etc., while chemical disruption can be carried out using enzymes, acids, and bases (Flanagan et al., 2008).

In the case of liquid samples, sample separation can be carried out using a variety of methods. Some of these methods include the following:

1. *Direct analysis after off-line treatment* which may involve filtration, dilution of sample with internal standard (for example, dilution of blood using aqueous internal standard for GC analysis), or protein precipitation with analysis of supernatant resulting after centrifugation.
2. *GC headspace analysis* which involves the establishment of equilibrium between the volatile components of a sample and the gas phase present above it, i.e., headspace.
3. *Liquid-liquid extraction* involves the separation of analyte from sample with a water immiscible and inert organic solvent at an appropriate pH.

Table 1.3 Sampling collection, preservation, and applications of some of the biological samples for toxicological analysis (Denis-Oliviera et al., 2017).

Sample type	Application	Sample collection and preservation
In vitro samples		
Exhaled air	• Useful for the analysis of volatile compounds such as ethanol and other xenobiotics such as cocaine.	
Oral fluid	• Useful for analysis of drugs of abuse. • Benefits include noninvasive sampling and less adulteration.	Collect 1–2 mL in a plastic container with preservative.
Breast milk	• Analysis of body burden of contaminants in adipose tissue. • Important route of exposure for xenobiotics in lactating females.	Collect 300 mL in a plastic container without preservative.
Sweat	• Useful for workplace testing • Noninvasive testing. • Collection requires time and is not easy.	Collected for one week.
Antemortem or postmortem samples		
Peripheral blood	• Offers complete toxicological analysis.	Collect 10 mL in a plastic container and add fluoride salt for preservation. For postmortem samples, collect femoral blood, while for antemortem, collect from cephalic vein.
Urine	• Higher concentration of xenobiotics and their metabolites as compared with other samples.	Collect 30 mL or all available in plastic containers.
Gastric content	• Collected for complete toxicological analysis.	Collect 30 mL aliquot in a plastic container with screw cap or 30 g of the stomach (without preservative).
Head hair	• Help establish a historical pattern of xenobiotic exposure.	Collect 150 to 200 hairs from posterior region of scalp. Store at room temperature in sealed envelops.

Continued

Table 1.3 Sampling collection, preservation, and applications of some of the biological samples for toxicological analysis (Denis-Oliviera et al., 2017).—*cont'd*

Sample type	Application	Sample collection and preservation
Finger and toe nails	• Useful in decomposed bodies where conventional samples cannot be obtained.	Collect nails from all fingers and toes using Teflon-coated stainless steel scissors.
Kidney	• Collected for the level of heavy metals or ethylene glycol analysis.	Collect 30 g in a plastic container with screw cap without preservative (postmortem analysis).

Table 1.4 Sampling procedures for some of the environmental toxicants (INTERPOL, 2014).

Environmental toxicant	Sample medium	Sample collection, storage, and holding time
Dioxins and furans	Biota, sediment, and industrial effluent	Collect via grab or composite sampling in 1-L amber glass bottle or in case of sediments, 180 g in wide mouth glass jar. Preserve at 4°C. Holding time = 30 days.
Pesticide	Water	Collect 1–4 L in amber glass containers. Preserve at 4°C. Holding time = 7 days.
	Biota	Collect one or more whole organism.
	Sediments	Collect 180 mL samples in amber glass containers. Preserve at 4°C. Holding time = 30 days.
Metals	Industrial effluents	Collect via grab or composite sampling in 250-mL high-density polyethylene containers. Preserve using nitric acid (2 mL of HNO_3 for 250 mL of sample).

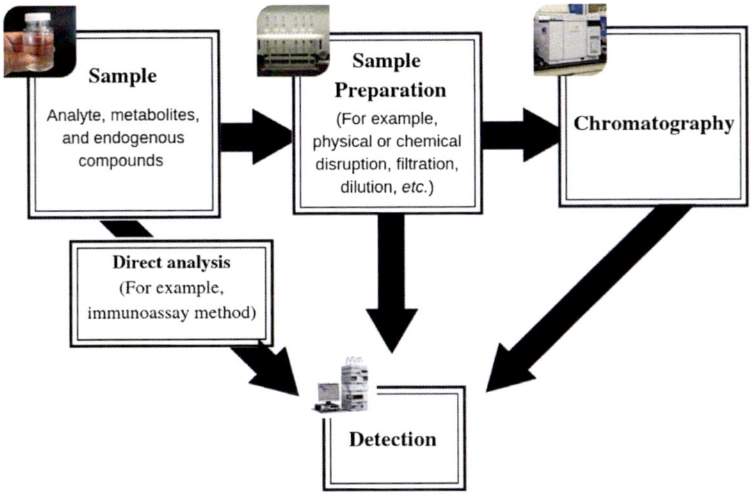

FIGURE 1.4 Schematic diagram of steps involved in sample preparation.

4. *Solid phase extraction* which involves the extraction of analyte from sample by adsorption onto solid materials such as ion-exchange resins followed by washing of sample and extraction of analyte (Flanagan et al., 2008).

The general methodology for the analysis of samples includes the following steps (Fig. 1.4):

1. *Detection* of drug or poison in the submitted sample by means of screening test. Some of the analytical techniques involved in screening of sample include immunoassay methods and thin layer chromatography.
2. *Identification* of drugs, metabolites, or poisons present by means of physiochemical test. Analytical techniques that can be used for this purpose include gas chromatography, high-performance liquid chromatography, infrared spectroscopy, ultraviolet-visible spectroscopy, GC-mass spectrometry, and HPLC-MS.
3. *Quantification* of drugs, metabolite, or poison.
4. *Interpretation* of results obtained from identification and quantification with reference to the information given and questions asked (Cooper and Negrusz, 2013).

Some of the analytical techniques that are most commonly used in analytical forensic investigations are tabulated in Table 1.5.

Table 1.5 Analytical techniques that are most commonly used in analytical forensic investigations (Flanagan et al., 2008; Cooper and Negrusz, 2013).

Analytical technique	Principle of technique	Applications	Nature of sample	Coupled techniques
Gas chromatography (GC)	Partition of sample in stationary and gaseous mobile phase based on retention time of the components.	• Analysis of drugs and organic poisons. • Crime scene testing. • Environmental analysis. • Forensic pathology. • Food industry.	Liquid and gaseous	GC-MS GC-IR GC-Flame Ionization Detector (GC-FID)
Mass spectrometry (MS)	Ionization of sample followed by the acceleration, deflection, and detection of ions.	• Clinical medicine and drugs.	Vaporized	GC-MS ICP-MS Ion mobility mass spectrometry Isotope radio mass spectrometry (IRMS)
Fourier transfer infrared (FTIR)	Involvement of IR with organic material and identification of chemicals based on functional groups.	• Agriculture. • Geology. • Biological samples. • Chemical fingerprinting.	Solid, liquid, and gas	ATR-IR IGI-R
Inductively coupled plasma atomic spectroscopy (ICP-AES)	Determination of the quantity of chemical in a sample from the intensity of light emitted after the atomization, excitation, and de-excitation of the sample.	• Agriculture. • Foods and beverages. • Biology • Geology • Water analysis and monitoring • Organics	Solid, liquid, and gas	ICP-MS
Atomic absorption spectroscopy	Determination of quantity of a chemical by the adsorption of radiation by free atoms in gaseous state.	• Agriculture • Foods and beverages • Biology • Geology • Water analysis and monitoring	Solid, liquid, and gas	ICP-AAS

UV-visible	Determination of concentration of a chemical in the sample by the measurement of attenuation in the electromagnetic beam after it passes through the sample.	• Waste water studies • PCBs • Phenols • Soil samples • Water quality analysis	Liquid Solids	Nil
High-performance liquid chromatography (HPLC)	Separation of the components of a sample mixture in mobile phase by interaction with stationary phase (alumina, C, Si) for the identification and quantification of individual components.	• Drug analysis • Water analysis • Biomonitoring • Urine analysis • Bio-sample • Anabolic steroids • Food and flavor industry	Liquid	HPLC-IR HPLC-MS
Inductively coupled plasma mass spectroscopy (ICP-MS)	Quantitative and qualitative analysis of sample after injection, atomization, ionization, fragmentation of sample.	• Bio-based applications • Earth sciences • Forensic sciences • Environmental sciences • Food industry • Archaeology	Liquid	ICP-AES
Immunoassay methods	Competition between antigen and labeled antigen for binding sites on an antibody raised against the antigen and determination of antigen concentration in the sample from proportionality of resulting signal.	• Drug screening		Radio immunoassay (RIA) Enzyme-linked immunosorbent assay (ELISA)
Thin layer chromatography (TLC)	Separation of substances on a planer surface (e.g., glass, aluminum, or plastic) coated with a solid phase (hydrated silica SiO_2)	• Pharmaceutical industry • Poison screening • Stability testing • Biological testing	Liquid	High-performance TLC (HPLC)

1.3 Case studies

1.3.1 Rapid population decline in oriental white-backed vulture (*Gyps bengalensis*) colonies due to diclofenac poisoning

Oriental white-backed vulture (*Gyps bengalensis*) was a common vulture species of the Indian subcontinent. However, since the 1990s, 99% decline in the population of oriental white-backed vultures has been observed in India, Pakistan, and Nepal. Owing to this large-scale and rapid decline in its population, it has been enlisted as a critically endangered species in the IUCN Red List of Threatened Species. Investigations have revealed that the rapid decline in its population was due to renal failure and visceral gout. When further research was carried out to determine the agent responsible for these symptoms via toxicological analysis, it was revealed that renal failure was a symptom of diclofenac poisoning resulting from the consumption of dead livestock containing the residues of this drug. Diclofenac is a nonsteroidal antiinflammatory drug used as a painkiller for livestock and has been associated with the decline of several vulture species (Oaks et al., 2004). For this purpose, it was banned across South Asia for veterinary use in 2006. However, some latest reports indicate that this drug is still being sold in Pakistan (Kazi, 2017).

1.3.2 Presence of aflatoxins in food items of Pakistan

Aflatoxins (AFs) are toxic compounds that are generated by some fungi found on agricultural crops such as wheat, maize, and ground nuts. These toxins are mutagenic, carcinogenic, and immunosuppressive in nature. Up till now, some 20 types of AFs have been identified, among which the well-known types are B_1, B_2, G_1, and G_2 (Mushtaq et al., 2012). Several research studies have been carried out in Pakistan to evaluate the occurrence of AFs in a variety of food items. Some of these studies have been tabulated in Table 1.6.

1.3.3 Imposex in marine gastropods as an indicator of organotin compounds in coastal waters of Pakistan

Organotin compounds, such as TBT and triphenyltin, were heavily relied upon in the past as antifouling agents in paints. Such antifouling paints were applied on ship hulls to prevent biofouling. However, since 2008, the

Table 1.6 Some of the studies conducted in Pakistan regarding presence of aflatoxins (AFs) in food products.

Objective of study	Employed analytical technique	Results obtained	Reference
Analysis of AF M_1 in milk samples collected from local markets in Karachi	ELISA	Mean value of AFM_1 in 91.7% samples was 346.2 ng/L which is higher than maximum limit prescribed by European Union (EU) (50 ng/L)	Asghar et al. (2018)
Occurrence of AFs in unpacked composites spices, such as biryani, nihari, korma, and tikka masala	HPLC	AF detected in 77% of samples of which only 3% contained AFs higher than the maximum limits set by the US	Asghar et al. (2016)
Evaluation of AFs in dry fruits and nuts from northern areas and Khyber Pakhtunkhwa, Pakistan	Reverse phase −HPLC (RP-HPLC) with fluorescence detector	13% and 24% samples found above permissible limit for total AFs and AFB_1	Masood et al. (2015)
Occurrence of AFs in 20 processed products from Pakistan (Cerelac, powder milk, biscuits, porridge, noodles, etc.)	RP-HPLC with fluorescence detector	21% of processed food items contained AFs in levels above those prescribed by EU (0.1 μg/kg)	Mushtaq et al. (2012)

use of these materials in antifouling paints has been banned worldwide owing to their deleterious effects on marine life such as imposex, teratogenicity, immunotoxicity, and neurotoxicity. Some of the main routes of entry in the marine environment include leaching of antifouling paints from ships and the dumping of ship breaking wastes in the seas (Hassan et al., 2014; Afsar et al., 2015).

Imposex is defined as the development of male characteristics in female animals and is a direct response of the occurrence of organotin compounds. Studies have revealed that this abnormality appears as a result of endocrine disruption, which interferes with the regulation of testosterone. In a study conducted by Hassan et al. (2014), the testosterone levels of two gastropod species were analyzed using radio immunoassay from Clifton, Manora Rocky Ledge and Gadani Ship-Breaking Yard. The results revealed that imposex was observed in all the samples from Gadani Ship-Breaking Yard, while no imposex was observed in the rest of the study areas. The higher level of testosterone in female snails at Gadani was attributed to release of TBT into marine environment by ship breaking and recycling process. In another study conducted by Afsar et al. (2015), data regarding imposex in *Thais carinifera* was recorded in six sites in Balochistan and Sindh from 1993 to 2012. A decline in imposex was observed from 2003–04 to 2012, which can be attributed to the global ban on organotin compounds.

References

Afsar, N., Siddiqui, G., Ayub, Z., 2015. Bioindicator *Thais carinifera* (mollusca, gastropoda): imposex response and consequences along the Pakistan coast during the period from 1993 to 2012. Brazilian Journal of Oceanography 63 (2), 115–124.

Altug, T., 2002. Introduction to Toxicology and Food. CRC press.

Asghar, M.A., Zahir, E., Rantilal, S., Ahmed, A., Iqbal, J., 2016. Aflatoxins in composite spices collected from local markets of Karachi, Pakistan. Food Additives and Contaminants: Part B 9 (2), 113–119.

Asghar, M.A., Ahmed, A., Asghar, M.A., 2018. Aflatoxin M1 in fresh milk collected from local markets of Karachi, Pakistan. Food Additives and Contaminants: Part B 1–8.

Cooper, G., Negrusz, A. (Eds.), 2013. Clarke's Analytical Forensic Toxicology. Pharmaceutical Press.

Dinis-Oliveira, RJ., 2017. Metabolic profile of oxazepam and related benzodiazepines: clinical and forensic aspects. Drug Metabolism Reviews 49 (4), 451–463.

Flanagan, R.J., Taylor, A.A., Watson, I.D., Whelpton, R., 2008. Fundamentals of Analytical Toxicology. John Wiley & Sons, West Sussex, SXW.

Frank, P., Ottoboni, M.A., 2011. The Dose Makes the Poison: A Plain-Language Guide to Toxicology. John Wiley & Sons.

Gupta, P.K., 2016. Fundamentals of Toxicology: Essential Concepts and Applications. Academic Press.

Hassan, S., Siddiqui, G., Saeed, S.A., &Sagheer, M., 2014. Influence of tributyltin (TBT) on level of free testosterone in two species of neogastropods found along the coast of Pakistan. International Journal of Sciences: Basic and Applied Research 14 (1).

INTERPOL, 2014. Pollution Crime Forensic Investigation Manual (Volume I of II). Retrieved from: https://www.interpol.int/Media/Files/Crime-areas/Environmental-crime/Leaflets-brochures/INTERPOL-Pollution-Crime-Forensic-Investigation-Manual-volume-1.

Kazi, M., 2017. Birds of Prey: Pakistan's Vultures Are Dying Out. Retrieved from: https://tribune.com.pk/story/1582412/1-birds-prey-pakistans-vultures-dying/.

Madsen, J.M., Traub, S.J., Burns, M.M., 2017. Chemical Terrorism: Rapid Recognition and Initial Medical Management. Waltham, MA.

Masood, M., Iqbal, S.Z., Asi, M.R., Malik, N., 2015. Natural occurrence of aflatoxins in dry fruits and edible nuts. Food Control 55, 62–65.

Mushtaq, M., Sultana, B., Anwar, F., Khan, M.Z., &Ashrafuzzaman, M., 2012. Occurrence of aflatoxins in selected processed foods from Pakistan. International Journal of Molecular Sciences 13 (7), 8324–8337.

Oaks, J.L., Gilbert, M., Virani, M.Z., Watson, R.T., Meteyer, C.U., Rideout, B.A., Mahmood, S., 2004. Diclofenac residues as the cause of vulture population decline in Pakistan. Nature 427 (6975), 630.

Pennington, D.W., Margni, M., Payet, J., Jolliet, O., 2006. Risk and regulatory hazard-based toxicological effect indicators in life-cycle assessment (LCA). Human and Ecological Risk Assessment 12 (3), 450–475.

Timbrell, J.A., 2008. Principles of Biochemical Toxicology. CRC Press.

Further reading

Chemical Risk Assessment: Overview and Examples, 2018. Retrieved from. http://www.chemsafetypro.com/Topics/CRA/introduction_to_chemical_risk_assessment_overview_principles.html.

CHAPTER

Forensics archaeology

2

Palwasha Gul, MPhil, Hareem, MPhil, Shazia Iftikhar, PhD

Department of Environmental Sciences, Fatima Jinnah Women University, Rawalpindi, Pakistan

Chapter outline

Abstract

This study is focused on the relation between forensics and archaeology, history of forensics archaeology, phases involved in field investigation, standards and guidance for an archaeologist, techniques and case studies related to Pakistan. Different

Trends of Environmental Forensics in Pakistan. https://doi.org/10.1016/B978-0-12-819436-2.00002-9

case studies related to Harappa and Mohenjo Daro have been discussed, that how the archaeologists build the relation between these two and called the Mohenjo Daro as one of the foremost settlements of Harappan Civilization. Last part of this chapter discusses the present condition and the future perspectives in context of Pakistan. The partnership of forensic science and archaeology in Pakistan can lead to a creation of very important scientific field that will support the investigation of crimes and terrorism. Most of the crimes remain untraced due to the lack of appropriate research and methodological expertise in the forensic science. So, especially in Pakistan, archaeological approaches are required to recuperate each and everything from the crime scene, with modern archaeological methods.

Keywords: Archaeological techniques; Archaeologists; Archaeology investigation; Archaeology: forensics archaeology: standards; Guidance.

Graphical Abstract

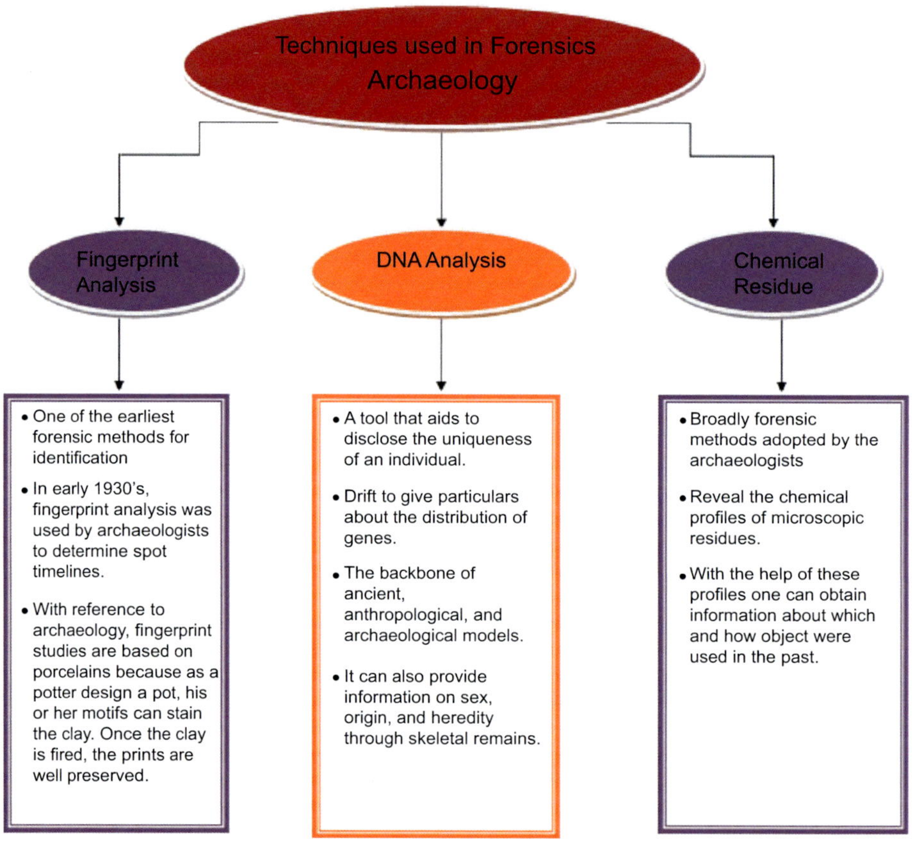

2.1 Introduction

Archaeology is the methodical study of past human societies by the examination and analysis of the items they left behind (Nawrocki, 1996). It provides essential ways to deal with buried remains and to assess and record a grave site and its surroundings (Crossland, 2013). Archaeology is a subdiscipline of anthropology, which is the study of science of humanity and humans in past or present. Geology and paleontology are the two disciplines that are often confused with archaeology. But archaeology is different and archaeologists are often trained in geology and paleontology and their center of concern is only humans and human remains.

There are some basic principles of archaeology that can be useful to forensic science. These principles consist of provenience, framework, structures, stratigraphy, superposition, taphonomy, etc. (Dupras et al., 2006). An archaeological digging and a crime scene are similar in many ways. Both disciplines inspect a past action and preserves past event by recording, assembling, preserving, and inferring physical evidence. Practices are used to dig archaeological interments with careful measurements and documentation. The precise scene of all objects in situ provides the source for reconstructing the crime scene (Haglund and Sorg, 2002). Due to the great resemblances between archaeology and forensic science, archaeological approaches can be applied to forensic science. The joint relation of archaeology and forensic science results in a new discipline called forensic archaeology.

Forensic archaeology is the use of archaeological and forensics methods for resolving the medical-legal problems. Recovery of human remains and other evidences from the forensic scenes is done by forensic archaeologists. Absolute and real archaeological process basically demands notable time and attention for detailed procedure. That is why this process seems rather to investigators. However, the result of this attempt will be the rebuilding of the exact whole scene as it looked before excavation.

With the help of the data and information given by forensic archaeologist, the inquisitor may be competent to do the following:

1. Justify or eliminate testimony from a witness or onlooker
2. Plot a link between a mugger and the scene
3. Recreate the situation of the decedent's death
4. Establish how and when the human remains turned out to be at the recovery site
5. Think about the natural factors that may have changed the condition of the scene after it was designed

6. Be sure with the evidences found at the scene that it is related to the case
7. Go to the court with the guarantee that all the work done in a scientific manner by the professionals

The term evidences used in archaeology includes all the components and things recovered from the site such as human remains, artifact, animals, plants and soil samples, footprints, tool marks, grave shafts, and fluid stains. Human remains include bones and soft tissues, including teeth, hair, and nails. Artifacts related particularly to the physical items that has been created, used, and manipulated by humans, but this may not be limited to only jewellery, coins, clothing, weapons, projectiles, and trash.

In some countries, such as Brazil, forensic archaeology was emerged as a branch of anthropology called physical anthropology, or biological anthropology, and in other countries, as a branch of archaeology. Archaeology provides a necessary set of methods for dealing with buried remains and for assessing and recording a grave site and its immediate surrounding environment.

2.2 History

We have been fascinated with the human body for a number of years, nearly as long as modern humans (*Homo sapiens sapiens*) have been around. But importance, specifically in human variation as a subject, of biological and physical anthropology emerged in the 15th century, when Europeans revealed and began colonizing the Americas. The indigenous populations wore different fashions, spoke foreign languages, built altered structures, and, most essentially, looked different. Thus, the Europeans attempted to classify the populations into groups which were based on physical and social qualities (height, features, figure, language, location). Unluckily, there was very little science in these early categorizations.

But, in the 19th century, science acquired an uncertain foothold in the study of human development and human diversity. Fossils were studied and investigated and measurements of living subjects were made, and in the publication of Charles Darwin's *On the Origin of Species*, it was discussed that how natural selection formed human (and other animal) evolution and development.

Forensic anthropology as it is also called physical anthropology or archaeology as we know it, however, made a distinctly late emergence to the biological anthropology party. Its history can be divided into four periods.

2.2.1 Formative period (1849–1938)

1. In Boston, Dr. John Webster murdered Dr. George Parkman in 1849 because he was to still to pay Parkman relatively a bit of money. However, two anatomists (Drs. Oliver Wendell Holmes and Jeffries Wyman) effectively reassembled the skeletal remains and resolute that they matched the stature, ancestry (Caucasian), and approximate age of Dr. Parkman
2. In 1859, *On the Origin of Species* is published.
3. Thomas Dwight gave his *Shattuck Lecture* about the analysis of human skeletal remains in a legal setting at Harvard in 1894, thus, he was justly called as the Father of Forensic Anthropology.
4. Two large, skeletal collections, Hamann-Todd (3000+) and Terry Collections (1600+), were created from 1900 to the 1920s. These collections were critical in building a basis for the fledgling forensic anthropology, they are no longer very useful. Stunted growth, porous bones, brittle bones showed that most of the skeletons belonged to individuals of low-socioeconomic status and showed evidence of poor physical condition and nourishment. In addition, as time goes further, humans as a species are emerging taller, which makes the early-20th-century skeletons quite dated.

2.2.2 Consolidation period (1939–71)

1. *The Human Skeleton in Forensic Medicine* was published in 1962 by Anthropologist Wilton Krogman. Forensic anthropologists and forensic archaeologist use this book even today.
2. After World War II, Hawaii founded JPAC (Joint POW/MIA Accounting Command) to locate position, identify, and then return the bodies of soldiers back to their families. The bodies were also found, identified, and returned soldiers from the Korean and Vietnam wars by JPAC. It is presently the leading human identification laboratory in the globe and one of the only two nationally certified human ID labs in the US.
3. Tom McKern and T. Dale Stewart published a report *Skeletal age changes in young American males* in 1954. In this report, McKern and Stewart established skeletal aging techniques mainly based on data from skeletal leftovers of Korean War soldiers.
4. In the early 1960s, the computer was invented, and this made the forensic anthropologists enable to use discriminant function analysis, a numerical technique, to identify sex and ancestry of skeletal remains.

2.2.3 Modern period (1972–99)

1. The American Academy of Forensic Sciences (AAFS) created a division for forensic anthropology called the Physical Anthropology Section.
2. In 1977, the American Board of Forensic Anthropology (ABFA) was formed to certify professional forensic anthropologists and to set professional standard.
3. In 1980, Dr. William Bass put up the "Body Farm" at the University of Tennessee, Knoxville, to investigate human decomposition and taphonomy.
4. In 1986, Dr. Richard Jantz and Stephen Ousley developed FORDISC, a computer program that allow forensic anthropologists to execute metric analyses on skeletal remains based on records from existing populations. The aim is to determine sex, ancestry, and height from the measurements.
5. At the same time, Dr. Clyde Snow began teaching sessions for forensic anthropologists to work on human rights and genocide cases. The first team was the Argentine Forensic Anthropology Team (EAAF).
6. Around that time, Dr. William R. Maples founded the C.A. Pound Human Identification Laboratory (CAPHIL), the earliest private forensic anthropology lab, at the University of Florida

2.2.4 Fourth era (2000-present)

1. The *Ellis Kerley Foundation* was established as the first financial support exclusively for forensic anthropological research in 2000.
2. In 2003, the first skeletal identification lab was JPAC to obtain ASCLD-LAB (American Society of Crime Laboratory Directors-Laboratory Accreditation Board) accreditation. CAPHIL followed shortly after.

2.3 Objectives of forensic archaeology

1. Discover and keep a track record of the context of the forensic scene.
2. Verify if the remains that are recovered at the scene are of human or nonhuman.
3. Make a list of the number of individuals buried.
4. Recognize whether the human remains are of prehistoric, historic, or forensic in nature.
5. Recover clues to find out about the manner of death.
6. Recover clues concerning what has occurred to the body and site after the death.

2.4 How forensic archaeology is different from traditional archaeology

In forensic archaeology, the site is made more intricate due to the presence of police officers, coroners, and media persons, whereas in traditional archaeology, the scenes may seldom be this complex.

1. Related deposits are usually completely hidden at prehistoric locations. Relevant deposits nearly consist of the current ground surface at forensic scenes such as the committal feature normally releases directly onto the current ground surface, and remnants may be dispersed extensively crossways the surface. So, it means that the ground used to get the site is a portion of the forensic scene and mere existence may amend or abolish evidence.
2. Forensic evidence is plentiful wider and possibly more refined than the artifacts archaeologists naturally come across. For example, the forensic archaeologist must be delicate to the existence of evidence such as cut roots, dry vegetations and dead flora, tool marks, shoeprints, and finger impressions.
3. The artifacts encountered at forensic scenes are often fragile and are seldom encountered at typical archaeological spots. These things may contain paper, fabric, tobacco products, pest evidence, hair, fingernails, and further soft tissues. The assemblage, packaging, cleaning, and protection of such evidence entail different drill and maintenance.
4. The opportunity of encountering biological, harmful, or hazardous things such as laden weaponries is more at forensic sites.
5. Measurable archaeologists should constantly administer a correct legitimate chain about authority for the collection of evidence, guaranteeing that specimens should be accounted to and ensured until they get submitted to police and different specialists. The greater part methods, procedures, notes, and discoveries might in the end be doubted previously, a court from claiming law, also others legitimate documentation

These exceptional qualities recommended that medicolegal authorities ought to utilize archaeologists who need preparation and experience in measurable investigations.

2.5 Phases of archaeological field investigations

There are three stages of forensic archaeological field investigations.

2.5.1 Phase I: systematic and comprehensive search/investigation (reconnaissance exploration)

1. The environmental clues in the area of study are also very essential in addition to the prior knowledge about the site under study to investigate the human remains. Local environmental circumstances, such as degree of visibility, terrain, etc., are very significant in the basic investigation technique and it more or less depends on the environmental circumstances. Burial and ground disturbance could be one clue in the variation or changes in the existing vegetation, soils, rocks, and insects.
2. Optical or pedestrian search: The comprehensive search of a region without creating disturbances to the objects.
3. GPS, aerial photography, thermal (infrared) imagery, and geophysical methods (e.g., ground penetrating radar, electrical resistivity, metal detectors) are all technological aids that might be helpful in the search.
4. For the identification of burials, decomposition (cadaver) dogs may aid, given the right climatic conditions.

2.5.2 Phase II: determination of forensic significance

1. Authenticate that the remnants are certainly *Homo sapiens* and not faunas.
2. Evaluate the age of the remnants (evaluate the postmortem interval).

The remnants may not be contemporary and scientific in character; medieval and premedieval may be muddled with the contemporary remains. Evidence pertaining to the age of the remnants comprises the condition of preservation of the remains (though this is determined by the local ecological situations), existence or absenteeism of contemporary dental (e.g., amalgams) or medical (e.g., surgical, pins, plates, or screws) care, and nature of related artifacts (e.g., funerary items) (Boyd, 2013).

Had the remnants be evaluated as premedieval or medieval, then the suitable historic conservation officer should immediately be referred before progressing (in Virginia Department of Historic Resources). If the remnants are on central terrain or administered by federal endowments, then NAGPRA (Native American Graves Protection and Repatriation Act) provisions (if the remnants are Native Americans). If the remnants are evaluated as scientific significant, suitable law implementation organization and medical inspectors or coroners should be informed, before progressing.

In certain cases, Step 1 of Phase II may be successful before Phase I is significant. If law application officers stumble upon or retrieve bone splinters they cannot recognize, they can direct photographs to a skilled scientific anthropologist, who can find out whether the bones are *H. sapiens* or not. If the bones can graphically be determined to be not human, then proceeding examination is removed.

2.5.3 Phase III: recovery; excavation

1. Record location landscape, weather, etc. through ground sketches, notes, and pictures. It is important to keep a photographic log.
2. After the identification of surface remains or an achievable grave site, photograph the site as it originally seems and then prudently eliminate vegetation without interrupting, possibly objects.
3. Ascertain datum and grid over examination site.
4. Start field map of site and map in place the place of all surface entity, photographing all the unprotected objects or remnants using a scale and north arrow

Surface remains will be treated as follows:

1. Find out if the remnants are determined and if the remains are in situ or dispersed. If dispersed, look for the area prudently for extra remains. Spot with inspector flags and map each place.
2. Monitor all soil in relation with the remains through ¼-inch net screen. Original location of remains is set up and approximation route of dispersion (based on landscape and slope, for example).
3. If not dispersed, a smaller grid can be founded to incorporate and map an individual body. Prudently examine the body site for teeth, as they can be critical in individual recognition. Screening of the soil around and beneath the cranium is necessary.

2.5.4 For buried remains

1. Outline limitations of the grave and map these in.
2. Describe the traits of the soil (color, texture) of the grave in contrast to the neighboring soils.
3. Split grave in half and prudently scratch away (with a trowel) the soil horizontally (*not* vertically), upholding a leveled surface and screening all unearthed soil. For intensely buried remains, digging of the soil in uninformed stages may be important. Once remains are uncovered, brushes and small (bone or wooden) tools are preferable.
4. Leave the place by pedestaling when bones or other unimportant objects are confronted.

5. After exposing the remains to the bottom of the grave and by taking soil samples in various areas, gather any proof of insect remains.
6. Record uncovered remnants by drawing and photography, documenting the characteristics of the photograph in the log book. Mapping includes both the construction of a two-dimensional plan view of the site as well as recording the vertical (depth) dimension of objects and remains in relation to the grid datum. Produce a profile drawing of the unprotected vertical half of the grave.
7. Dig up the remaining half of the burial in a similar manner.
8. Measure the ultimate depths of bones/remains and the bottom of the grave from the datum, and record these in the log book.
9. Recognize individual artifacts, bones, or body parts numerically on turf illustrations, store them in separate bags, placing the corresponding numbers on these bags.
10. Backfill the unearthed grave with the screened mud.

Documentation of all stages with field records, drawings, and snapshots is vigorous for recreating the scene. The usage of technologically sophisticated apparatus (Total Stations, LiDAR) is extremely suggested if it is accessible specifically for intricate scenes with several stuffs to plot and record. This will help to document the procurement of various sources of information to boost the understanding of the scene and the proceedings involved in its formation.

2.6 Standards and guidance for archaeologist

1. Archaeologists necessarily have a basic acquaintance of law implementation and authorized trials
2. Must be talented to collaborate with law implementation personnel
3. Must be able to demean our inquiries under severe time limitations and mass media consideration
4. Must be able to pact with circumstances that fluctuate from conservative archaeology such as burials including soft tissue

2.7 Techniques

2.7.1 Fingerprint analysis

Fingerprint analysis is one of the earliest forensic methods for identification. Fingerprint analysis has made the archaeologists in this field for a number of years. With reference to archaeology, fingerprint studies are based on porcelains because as a potter designs a pot, his or her motifs can stain the clay. Once the clay is fired, the prints are well preserved. In the early 1930s, fingerprint analysis was used by archaeologists to determine spot timelines. Dr. William F. Bade used fingerprint technique in Tell en-Nasbeh site in Palestine to understand the complicated artifact deposits at the site. After careful investigation, he concluded that one potter had engraved numerous vessels and so any layer that contained this potter's work belonged to the same timeline.

Fingerprint analysis also contributes in providing methods to recognize children by means of artifact analysis. Diggings at prehistoric Sinagua sites in the Southwestern United States have exposed small clay animal creatures from the interments of children. Most of them were clumpy irregular and uneven, these porcelains seemed to be the effort of those potters who were possibly at their learning stage, most probably children. The unfinished surfaces marked with fingerprints and nail impressions made them perfect artifacts for the analysis of fingerprints and age.

The studies conducted by Kathryn Kamp and her colleagues on modern populations have revealed that the age and width of fingerprints ridges are correlated. They analyzed both, the fingerprints from the animal figurines and the pottery produced by experts from Sinagua site. This resulted in making the idea of average age of figurine makers which was 11–13 years, whereas the average age of experts was 18 years or above. This research appeared fascinating because it allowed others to see how the prehistoric children were connected with their surroundings. These figurines showed that how they were learning about the properties of clay, to become expert potter.

2.7.2 DNA testing

Mostly, DNA analysis is a tool that aids to disclose the uniqueness of an individual. In prehistoric DNA studies, the drift to give particulars about the distribution of genes rendering to supposed forms of activities of specific peoples and societies is justly public. Ultimately, these traditions were the backbone of ancient, anthropological, and archaeological models. The archaeologists had a clue of the person they were observing for, so they were clever to excerpt a sample of DNA from his teeth and link it to the DNA present in his identified alive relations. DNA analysis can be used for plentiful further than positive proof of identity. It can also provide information on sex, origin, and heredity. Nuclear and mitochondria are the two common types of DNA. Nuclear DNA contains genetic information from parents, and mitochondrial DNA contains information from one's mother and is stored in specialized parts of cells. Hundreds of mtDNA are present in each cell. Both types can be analyzed but the type of analysis used depends on age and preservation of remains as well as information of interest of archaeologist.

From skeletal remains, thousands of years old nuclear DNA has been recovered and can help in the determination of an individual's sex and, at times, family background. Interestingly, nuclear DNA can also be recovered from artifacts that would have come into interaction with DNA-rich body fluids, such as saliva. In mid-19th century, archaeologists obtain DNA by this method while working on the site of Boston Saloon, in the mining town of Virginia City, Nevada. The archaeological site revealed flasks, clay plumbing, glassware, and porcelains. Unfortunately, these kinds of artifacts are especially tough to understand because they are not linked with a specific sex, age, or ethnicity. Due to this purpose archaeologists decided to use nuclear DNA analysis as an effort to unveil the sex of the individuals who used the stuffs. DNA recovered from a clay pipeline stem by different scientists exposed prominent teeth marks as evidence that it had been used. DNA rapidly destroys when visible to light, so the shaft from the pipeline stem provided an exceptional shielding environment. The consequence of the examination was astounding. The DNA collected from the shaft showed that a woman had used the pipeline. This evidence dissipates the statement that the saloon was a male-only environment.

Numerous archaeological locations are much older than the Boston Saloon site, and this is where mtDNA comes into play. mtDNA is more likely to survive integral in earliest remains because it is present in higher numbers than nuclear DNA. Even though mtDNA was used to find Richard III's 530-year-old bones, it has been used on remains significantly older. Recently, mtDNA was used for the determination of the human beginning of petrified coprolites unearthed from Paisley Cave in Oregon. The coprolites were out-of-date to an incredible 14,000 years ago. It is well accepted that humans occupied portions of the Americas as early as 11,500 years ago, but there has been intense dispute among archaeologists as to whether they were existing before this time. Some have faith in that the initial humans arrived when glaciers melted, forming an ice-free path through Canada into the Americas. This immigration would have happened no previous than 13,000 years ago. Other archaeologists sense that humans arrived the Americas much earlier and the human coprolites from Paisley Cave deliver some of the solidest evidence to date for this site. This early date for the coprolites also recommends that the first humans might have inwards in the Americas not on foot through the ice-free passage, but possibly along the coastline in yachts.

2.7.3 Chemical residue analysis

The utmost broadly forensic method adopted by the archaeologists is the chemical residue analysis. Before the use of this technique there was no idea about which vessel and stone tool was used for what, but now with the use of this technique one can reveal the chemical profiles of microscopic residues. With the help of these profiles one can obtain information about which and how objects were used in the past.

In 2008, chemical residue analysis assisted archaeologists to resolve one of the permanent mysteries of the ancient American Southwest. Pueblo Bonito is an overwhelming, ancestral Puebloan site in Chaco Canyon dating to AD 860–1128. Built of thick granite walls and huge timbers, the pueblo contained at least 800 rooms. In one of the rooms were found 111 cylinder jars tinted with prominent geometric black and white designs symbolic of the time. These vessels are very infrequent and still they possibly influenced a ritual worth, their exact use was unfamiliar. Archaeologists decided to tackle this mystery by the use of chemical residue analysis by examining the organic remains on wreckages of cylinder pots from an ancient trash deposit linked to Pueblo Bonito. The chemical profiles exhibited that the containers contained some remains of cacao. This result is important because it brings to light the possible use of the chamber containers and supports the presence of occupation and ceremonial associations between Chaco Canyon and Mesoamerican ancient cultures. Demanding a tropical climate of high rainfall and high temperatures, cacao could not be grown up in the setting of the ancient American Southwest, and therefore Pueblo Bonito was getting its cacao through trade with Mesoamerican civilizations. It is fine acknowledged that cacao brews were used in Mesoamerican ceremonies where the drink was fermented by pouring it from one container to another. The similar sort of ceremonial could have been going on at Pueblo Bonito, telling that the previous beliefs in Mesoamerica and the Southwest were not simply trading quantifiable belongings but also concepts.

2.8 Pakistan and forensic archaeology: case studies

2.8.1 Harappa

Harappan Civilization (or Indus Valley Civilization [IVC]) was named after Harappa village in Pakistan, the first archaeological site ever stated of the same civilization. It was discovered by Sir John Marshall and Daya Ram Sahani in 1921–22. In 1927, a cemetery site called as Cemetery H came into being due to natural weathering. Later in 1987–88, K.N. Sastri also revealed the Cemetery R-37 at the same location (Lal et al., 2003). Cemetery H (Late Harappan period) is situated nearby to the northwest area of Archaeological Museum of Harappa. Cemetery H interments can be divided into two groups. First, more than 135 of pot burials were found by the archaeologists in this area. This sort of burial was used for the departed babies (Sharma, A. K. 1999). Skulls were touched one side of the pot wall while the long bones were placed horizontally, crossing each other in several positions. These were primarily secondary fractional burials comprising only a few bones from a single individual or sometimes from numerous individuals. These pot burials were very famous in agricultural society of these ancient sites and finding of these at large scales may suggest that agriculturalists might have exist in these two ancient sites (Basit and Rahman, 2015) (Dibyopama et al., 2015). Next was primary pit burial that was common amid Harappan people buried in Cemetery H. The alignment of the skeletons of the primary pit burial was northeast to southwest or somctimes from west to east. Skeletons were usually laid in a side position with inflexed or extended legs.

Only one skeleton was laid in a prostrate position. In most of the cases, arms were bending and hands were placed in front of the face, though an exemption was revealed in the case whose forearms crossed each other at the stomach. In some of the cases, along with the human bones animal bones were also found. Burial goods were of humble nature such as long necked water vessel, offering dish, and flasks, etc. (Lal et al., 2003). On contrary to this, as for Cemetery R-37, total 90 burial remains were excavated from the site. The pits were mostly rectangular and the skeletons inside are oriented in north south direction. Maximum burials were stretched with supine position of the body. Heads were usually placed toward north while legs were toward south. Bricks lined some of the graves, whereas red-colored wooden coffins were found in the other graves. Stoneware was discovered as burial goods in every single grave. Away from potteries, shell bracelets, copper ornaments, steatite disc beads, carnelian beads, lapis, jaspers, microbeads are also found from the graves mostly along with female skeletons.

2.8.2 Mohenjo Daro

Mohenjo Daro is one of the foremost settlements of Harappan Civilization ever revealed. It is an archaeological place present in the province of Sindh, Pakistan. This site was discovered by Rakhaldas Bandyopadhyay as early as in 1922. In the 1930s, major diggings were additionally carried out at the site under the leadership of Sir John Marshall, D.K. Dikshit, and Earnest Mackay.

The final excavations had been finished in 1964 and 1965 (Rao, 1985). In contrast to its regal urban view, the well-disciplined cemetery from Mohenjo Daro has not yet been sited. As a substitute, the utmost notable and debatable feature in Mohenjo Daro was many skeletal remains dispersed in disorder throughout the prehistoric city. In fact, it is eccentric that many skeletons were found in the street of Mohenjo Daro, but not being buried in any other places there. Mortimer Wheeler thus related the scene with the so-called Aryan raid in the Sindh region, a kind of catastrophe initiating the demolition of Harappan civilization (Bisht, 1991). However, the consequent physical anthropological study by Kenneth Kennedy has verified that they were not murdered, but were died of some other tragedies much later than Harappan period (Nath, 1998).

2.9 The present condition and the future perspectives

In 1930, the first forensic laboratory was established in Lahore. By 1947, the laboratory was working as a training center and dealing with the investigation of forensic evidence (Ahmad, 1961). But today there are a few well-equipped forensic laboratories present in Pakistan. Archaeology is not a developed field in Pakistan as compared with the field of forensic science. And even in this modern scientific time, Pakistani archaeologists are using old traditional methods for archaeological investigations and examinations. The absence of interdisciplinary approaches and associations among the sectors of forensic science and archaeology are the reasons for the lack of forensic archaeology in Pakistan. There is an inadequate education of archaeology for forensic experts and vice versa. There is no single course of forensic archaeology offered in any university or institute of Pakistan. There are no chances for Pakistani archaeologists to be skilled in forensic science. The current condition of archaeology of Pakistan demonstrates that the status of forensic archaeology, but its importance has not yet been realized.

In future, the partnership of forensic science and archaeology in Pakistan can lead to a creation of very important scientific field that will support in the investigation of crimes and terrorism. Most of the crimes remain untraced due to the lack of appropriate research and methodological expertise in the forensic science. In most cases, dead bodies are buried in the ground by the assassins in order to hide their crime. So, especially in Pakistan, archaeological approaches are required to recuperate each and everything from the crime scene, with modern archaeological methods.

References

Ahmad, N.R., 1961. Our Police Heritage. WAPDA Printing Press, Lahore.

Basit, A., Rahman, G., 2015. Archaeology and forensic science: the need for an interdisciplinary approach in Pakistan. Pakistan Heritage 7, 1–10.

Bisht, R.S., 1991. Dholavira: new horizons of the Indus civilization. Purattatva 20, 71–82.

Boyd, C.C., 2013. Forensic Archaeology: Search and Recovery Best Practices.

Crossland, Z., 2013. Evidential regimes of forensic archaeology. Annual Review of Anthropology 42, 121–137.

Dibyopama, A., Kim, Y.J., Oh, C.S., Shin, D.H., Shinde, V., 2015. Human skeletal remains from ancient burial sites in India: with special reference to Harappan civilization. Korean Journal of Physical Anthropologists 28 (1), 1–9.

Dupras, T.L., Schultz, J.J., Wheeler, S.M., Williams, L.J., 2006. Forensic Recovery of Human Remains: Archaeological Approaches. CRC Press, New York.

Haglund, W.D., Sorg, M.H., 2002. Advances in Forensic Taphonomy: Methods Theory and Archaeological Perspectives. CRC Press, New York.

Lal, B.B., Joshi, J.P., Thapar, B.K., 2003. Excavations at Kalibangan: The Early Harappans (1960-1969), first ed. Archaeological Survey of India, New Delhi.

Nath, A., 1998. Rakhigarhi: a Harappan metropolis in the Saraswati-Drishadvati divide. Puratattva 28, 39–46.

Nawrocki, S.P., 1996. An Outline for Forensic Archaeology.

Rao, S.R., 1985. Lothal, New Delhi: The Director General of Archaeological Survey of India, India.

Sharma, A.K., 1999. The Departed Harappans of Kalibangan. Sundeep Prakashan, New Delhi.

CHAPTER

Pedospheric environmental forensics aspects

3

Shaan Bibi Jaffri, PhD Scholar, Arooba Nosheen, MPhil Scholar, Shazia Iftikhar, PhD, Khuram Shahzad Ahmad, PhD

Department of Environmental Sciences, Fatima Jinnah Women University, Rawalpindi, Pakistan

Chapter outline

Abstract

Lithosphere constitutes the most important compartment of environment. Soil microbes are essential for plant nutrition and health and play an important role in agroecosystem functioning. They have most diverse communities that reflect soil

Trends of Environmental Forensics in Pakistan. https://doi.org/10.1016/B978-0-12-819436-2.00003-0

type, environmental conditions, and land management. Environmental forensics possesses an important aspect for soil analysis for detection of different microbes found primarily in the pedospheric zone. There are different methods for characterizing microbial communities in soil, including genotypic and phenotypic analysis. The first forensic lab in Pakistan was established before partitioning, in 1906. A noteworthy issue in soil microbial examination has been that most soil microorganisms cannot be characterized by established microbiological cultivation techniques. It is evaluated that around 80%–90% of soil microorganisms are not yet cultured by methods of traditional strategies.

Keywords: Ecological niche; Geography; Microbial communities; Remote sensing; Soil profile.

Graphical Abstract

3.1 Introduction

Environmental forensic analysis and related domains are inclusive of all the environmental compartments since they are submerged into one another. In this regard, the significance of hydrospheric and atmospheric compartments cannot be overseen, but lithospheric compartment has always been the most targeted one. This can be explained by the receptive nature of soils for different types of contaminants. As in the normal biogeochemical cycles and ecological activities, soils are marked as recipient of different products in the form of some useful and majorly harmful compounds, e.g., agrochemicals and pesticides; in the same way, soils are also exposed to the material deposited from criminal activities. Furthermore, efforts have been done to replace the chemical agricultural products and replace them with biopesticides or advanced nanobiopesticides synthesized (Jaffri and Ahmad, 2017, 2018a-d), even then chemical spraying received by the soils has been tremendously increasing. In this regard, the microbes have developed resistance, needing advanced modes for abatement (Ahmad and Jaffri, 2018a,b). This is a relatively advanced field since the routine analysis are often based on the pedospheric examinations for agricultural, environmental, and geological investigation, but unlike these, forensic soil analysis aims at the correct identification of crimes by utilization of the exposed soils as an evidence. Soils can provide significant information on the type of activity, time duration, distance; and biological data can also reveal the type of casualty done as a result of any particular crime. Thus, forensic soil analysis finds its future prospects in the accurate crime identification by combining various physical, chemical, biological aspects with the software-based approaches. This chapter has been written to give an overview on the forensic soil analysis so that future researchers in Pakistan and other countries can utilize the knowledge in reaching the truth regarding any criminal activity with accuracy using pedospheric samples.

3.2 Pedospheric samples characterization

Soil samples possess a special significance for forensic purposes and criminal activity detection. Soil acts as a recipient medium for different materials that might be produced from any criminal activity. Soil is often tested for different physical, chemical, and biological makeup for detection of target materials. Such characterization modes are usually inclusive of the soil structural, visual, consistency, textural, chemical, and biological aspects. Such analytical evaluation of soils enables the scientific community for interpretation of ecological functioning and the role of soil in it. Pakistani soils in this regard have been examined for various parameters by different researchers (Tables 3.1 and 3.2) and a great deal of variability has been exhibited by different soils. In addition to the use of these aspects for forensic investigation, soil characterization is often done for carrying out the environmental conservation, management, and determination of the soil's health (Kheyrodin, 2014). However, forensic investigation based on soils is somewhat different from those done for monitoring purposes. Soil-based investigation

Table 3.1 Pakistan's soils unique physicochemical makeup expressed by different research studies.

Sr. No.	Region	Province	Major finding	References
1.	Abbottabad	KPK	Sandy loam with 0.70% organic matter, expressed adsorptive nature toward herbicide alachlor,	Ahmad (2018a)
2.	Taxila	Punjab	Clay loam with 5.62% organic matter, expressed adsorptive nature toward herbicide alachlor,	Ahmad (2018b)
3.	Warsak Dir	KPK	Selected soils expressed an organic matter of 7.45% and 4.60%, respectively. Soils' Cd sorption was enhanced by amendments,	Khan et al. (2018b)
4.	Kashmir	Azad Jammu and Kashmir	Kashmiri soils have an enhanced chance of deterioration by various process, e.g., physicochemical and biologically driven.	Abbasi et al. (2018)
5.	Kot Murad	Punjab	Soil amendment with different organic and inorganic compounds, substantially increased soil fertility, and augmented the soil health.	Mehdi et al. (2018)
6.	Rawalpindi	Punjab	Heavy metals contaminated soils can be phytoremediated and the extent of phytoremediation can be enhanced by fungal inoculation.	Ifthikhar et al. (2018)
7	Faisalabad	Punjab	Wastewater-irrigated soils comprise fungal cultures that are tolerant toward many heavy metals.	Iram et al. (2012)
8.	Lahore	Punjab	Pakistani soils contain a rich diversity of microbial species in comparison with German mine soils.	Iram et al. (2009)

Table 3.2 Soil microbes isolated from different soils of Pakistan.

Sr. No.	Region	Isolated microbes	Reference
1.	Lahore	• *Acremonium* sp. • *Alternaria* sp. • *Aspergillus niger* • *Aspergillus* sp. • *Aspergillus nodulans* • *Aureobasidium* sp. • *Chaetomium* sp. • *Coniothyrium* sp. • *Curvularia* sp. • *Fusarium* sp. • *Humicola* sp. • *Monilia* sp. • *Monocillium* sp. • *Mortierella* sp.	Iram et al. (2009)
2.	Faisalabad	• *Klebsiella* sp. • *Pseudomonas putida* • *Aeromonas* sp.	Ajaz et al. (2005)
3.	Lahore	• *Proteobacteria* • *Actinobacteria* • *Acidobacteria*	Sheik et al. (2012)
4.	Faisalabad	• *Rhizobium* • *Bradyrhizobium* • *Agrobacterium*	Hameed et al. (2004)
5.	Attock	• *Azospirillum*	Ilyas and Bano (2010)

is not only inclusive of the surficial contaminant determination but also involves the detection of the contaminant in the deeper lithospheric regions as well.

The deeper regions provide a better clue regarding contaminant volumes in comparison to the surface layers. The thorough comprehension of the soil depth and distribution of the contaminant in a vertical manner is significant in soil pollution estimation. Such an approach is also marked by the elucidation of the different depositional procedures in the lithospheric region. As the knowledge of the soil surface prior to organic amendments is a crucial step, so is the understanding of the pollutant distribution in the vertical manner for better managerial modes. Soils were characterized via USDA-NRCS pedon method which is expressed to be the best procedure for obtaining comprehensive knowledge on the variability in the soil depth (Morris et al., 2010).

Various studies have confirmed the role of soil erosion in the transferal of various contaminants. Soil erosion is often characterized for casting serious impacts on the higher slopes as compared with the lower or midslope regions. In this way, the soil mineral content is altered consequently altering the long-term productivity of soil (Khan et al., 2013). The pedospheric forensic examination of soils can be done by the soil scientists who have obtained expertise in the soil analysis for different parameters, e.g., classification, particle characteristics, climatological and hydrological influence, vegetation aspects, structural features, color, etc. The pedosphere forensic examination of soils is a necessary step for criminal investigation in case of Pakistan since the country has been used for various terrorist activities, killing a large number of innocent people and destroying natural resources. In addition to the destruction of Pakistani soils due to terror-based activities, there has also been a considerable damage caused due to unplanned urbanization. Such urban sprawls could not be provided with all the facilities as expected by the people migrating toward the cities and in turn their migration adds up to the already disturbed soils by various developmental activities. The following section throws light on the different characterization methods done for identification of soils.

3.2.1 Electrical resistivity

The electrical resistivity of soils could reveal a significant data for different forensic investigation purposes. Investigating the electrical resistivity of the soils is a nondestructive procedure that does not incur any disturbance in the soils under examination. Furthermore, electrical resistivity is also one of the favorable tools for delineation of the subsurface characteristics of the soils. Correlational analysis based on the electrical resistivity and soil physicochemical properties enable the subsurface soil characterization without the need for sampling via borehole. There is a considerable association between electrical resistivity and various parameters, e.g., plasticity index, internal friction angle, and moisture. Additionally, the water table in the deeper soil profiles influences the correlation models and yields considerably lower values (Siddiqui and Osman, 2012). Electrical resistivity has been expressed as a good in situ evaluation tool for the prediction of soil characteristics (Siddiqui and Osman, 2013).

3.2.2 Microbial community composition

Soil microbial composition is also crucial for complete forensic investigation. The typical type of endemic factors that are associated with the typical soils can reveal greater potential for investigations. For instance, the soil samples having similar types of microbial community can be used for derivation of the results. However, prior to the microbial composition analysis, it is necessary to understand that soil microbes' survival is dependent upon a myriad of factors such as pH, organic carbon quantitative and qualitative aspects, amount of nutrients, moisture presence, and oxygen levels. Microbial population's presence can be detected in all profiles of the soils and are associated with nutrient recycling and formation of the soils. Microbial populations' presence in different soil profiles is a challenging task; however, it is also marked as one of the significant criteria for successful forensic investigations. Researchers have reported the characterization of soil microbial communities and the alterations in various bacterial and archaeal populations in nine soils by utilization of the barcoded pyrosequencing of the 16S rRNA gene (Eilers et al., 2012).

3.2.3 Carbon pools

Soils are the sinks for the carbon on earth. A greater amount of carbon present in the terrestrial ambiance becomes part of soils. Soil carbon content determination is significant because present-day environmental forensics is aimed at global warming, and thus consideration of soils in this regard is a crucial factor. Although the carbon content of soils possess greater significance in comparison to the plants, however, the amount of data expressing the above ground data for carbon is scanty. The frequently done shallow soil sampling is usually marked by the underestimation of the soil carbon in different soil layers, and also there are inadequate results regarding soil carbon alterations as a result of different activities, e.g., fertilization, vegetation management, and tillage (Harrison et al., 2011).

3.2.4 Structural and visual characterization

Soil morphological features also provide a considerable clue for forensic investigations since soils possess, like all other geological objects, a three dimensional complicated structure. Structural features involve the spatial distribution and pore space in addition to the mechanical and hydraulic properties. The structural features of soils can be analyzed via tomographic techniques at a wide range of spatial dimensions. There are reports on the derivation of the various information regarding porosity, aggregates, surface area, and pore topology (Vogel et al., 2010). Soil color is often developed due to the presence of organic matter, moisture content, and the iron mineralogy (Moritsuka et al., 2014).

3.3 Current status of institutes carrying out forensic examination in Pakistan

The principal criminological lab in the area originated before the partition of India. It was set up in 1906 at the photographic area of Punjab's criminal examination office on Birdwood Road, Lahore. By 1947, the lab was currently filling in as an instructional hub and also managing examination of guns, fabric filaments, fake coins, manufactured money, mystery inks, and manually written and typed materials. After partition, the administration of Pakistan built up three more scientific science labs in Quetta, Peshawar, and Karachi. With passage of time, testimony and concessions have become the most preferable method for social affairs confirmation while criminological based examination philosophies became less favorable options. But the pattern took a turn after 2001. The government chose to fabricate more labs and redesign existing ones over the nation. A complete 10-year improvement design was intended for the reason. The labs were intended to help enhance criminal ID offices and upgrade the limit of law implementation organizations, through utilization of present-day logical apparatuses and most recent criminal examination procedures.

3.3.1 The National Forensic Science Agency

The National Forensic Science Agency (NFSA) was affirmed by the Executive Committee of the National Economic Council (ECNEC) in 2002, with an underlying spending plan of PKR 1.29 billion. The organization was set up as a self-ruling body to build up an all-inclusive strategy for advancement of scientific science research centers over the nation. A training establishment was likewise to be set up to create exceedingly talented specialized staff and legal specialists. Experts at the NFSA established a forensic lab which became practical in 2006; however, the arrangement to set up a preparation and showing establishment is yet to be actualized. Up until this point, the office has built up a unique fingerprint and DNA database with records of an excess of 200,000 offenders. The office has illuminated 2540 cases out of 2659 conveyed to it. That is a win rate of around 96%. The apparatus available with the NFSA incorporates genetic analyzers, centrifuge and PCR machines, a biosafety unit, an RT-PCR, vertical chillers, and workstations for the DNA segment. The office has additionally procured crime scene kits, emergency treatment units, and computerized cameras; brushes, tools, and Polilight to identify fingerprints; Regula 1010 and 1019; VSC 6000/HS; ESDA 2; stereo magnifying lens (Lica); and examination magnifying lens.

NFSA has 20 scientists who have been trained by Australian specialists in DNA, firearms, tool marks, fingerprint, ballistics, explosives, document analysis, and crime scene investigation. NFSA also has two other scientists who have been trained in DNA techniques in China and the UK, and one got training in narcotics testing from France. NFSA is not only a criminological laboratory but also an organization which has, so far, lots of skilled officials from Sindh, Balochistan, and Khyber Pakhtunkhwa (KP).

3.3.2 The Punjab Forensic Science Agency

The installed technology in Punjab Forensic Science Agency (PSFA) consists of triple-quadrupole mass spectrometer, gas chromatograph with flame ionization detector, gas chromatograph with mass spectrometry, gas chromatograph with NPD, liquid chromatography-mass spectrometry, and enzyme-linked immunosorbent assay. In the narcotics and chemistry lab, PFSA is using ultraviolet-visible spectrophotometry, FTIR, gas chromatograph with flame ionization detector, and gas chromatograph with mass spectrometry, The PFSA now has 14 diverse completely furnished criminological laboratories. These include labs specializing in DNA, crime and death scene investigations, serology, latent finger marks, firearms/toolmarks, forensic histology and pathology, narcotics, postmortem toxicology, question documents, chemistry trace, cybercrime investigations, audio visuals, and computer forensics. PFSA have so far solved 248,000 cases including kidnapping, robberies, murders, and incidents of terrorism. The PFSA is now ranked as the second largest forensic establishment of world. The PFSA has solved nearby cases as well as imparted its skill to distant nations. It went about as a referee lab in the Ohio police shooting in the US in 2015. PFSA's discoveries on spent projectiles and cartridges were approved in the US court as master and definitive ballistic investigation. The prosecutors had connected the PFSA after forensic exams at two labs in the Ohio state yielded inconsistent outcomes. PFSA has likewise given administrations to law requirement offices in the UK, Egypt, Saudi Arabia, and Norway.

3.3.3 Forensic science in Khyber Pakhtunkhwa

In the current past, KP police have built up an Institute of Forensic Science (IFS) in Peshawar with help from the United Nations Development Program (UNDP) and the Australian Federal Police. The organization now houses the Forensic Science Laboratory built up in 1976. The laboratory comprises 52 officials including 24 forensic experts. So far the laboratory has dealt with more than 40,000 cases. It is the vital crime examination lab in the territory, comprising six specific units for investigation of chemicals and narcotics, fingerprints, guns and tool marks, questionable documents, criminological photography, and advanced legal forensics. The lab gives criminological administrations to the Pakistan Customs, the Excise and Taxation Department, the Anti-Narcotics Force, the National Accountability Bureau, the Anti-Corruption Establishment, the Frontier Corps, the FATA organization, and Malakand Levies. Suggestions for upgradation of the laboratory and expansion of an independent forensic science agency are under deliberation.

3.3.4 Forensic science in Sindh and Balochistan

The first forensic laboratory in Sindh was established in 1958. This was trailed by the foundation of a Criminalistics Division in 1983 as the prepartition Fingerprint Bureau was converged with the Forensic Science Laboratory. The division was renamed as the Forensic Division in 2009. From that point forward, the legal wing of Sindh police has taken a shot at around 50,000 cases. A majority share of these cases were identified with target executing. The scientific division directly comprises units for audiovisual examination, computerized legal sciences, chemical monitoring, offense scene examination, guns and ballistics, forensic photography, robotized unique mark recognizable proof, and dormant fingerprint. In Balochistan, a provincial forensic laboratory was set up in the Police Department amid 1983–84. The laboratory has just two confirmed compound analysts. The available apparatus is of very low quality. The region's measurable science lab has just four segments. These are chemical, gun and toolmark, faulty archives, and fingerprints areas.

3.4 Soil microbial communities of human decomposition on soil

Soil microbes are essential for plant nutrition and health and play an important role in agroecosystem functioning. They contribute to global elemental cycling on a larger scale. In addition, they are involved in turnover strategies of organic matter, breakdown of xenobiotics, and organization of soil aggregates. As opposed to plant decent variety, and the macro- and mesofauna, the measure of soil microbial assorted variety is a somewhat new approach. A noteworthy issue in soil microbial examination has been that most soil microorganisms cannot be characterized by established microbiological cultivation techniques. It is evaluated that around 80%–90% of soil microorganisms are not yet cultured by methods for traditional strategies

(Schroder, 2008). The most various and assorted individuals from the soil food web are microorganisms—the bacteria and fungi—however, there are likewise numerous specie types of variable sizes that live in soil, including the microfauna (body width 0.1 mm, e.g., protozoa and nematodes), the mesofauna (body width 0.1–2.0 mm, e.g., microarthropods and enchytraeids), and the macrofauna (body width 2 mm, e.g., earthworms, termites, and millipedes) (Bardgett, 2005).

The soil microbial community reflects soil type, environmental conditions, and land management (Gattinger et al., 2008). The huge microbial range found in soils and their utilities and compliance to changing conditions make them appealing biomarkers of utilization in giving both search aptitude and possibly as evidence. In addition, the microbial profile can be attained from very small samples (<0.2g) (Sensabaugh, 2009). Despite the fact that the likelihood of applying molecular microbial methods to soil evidence for criminological purposes is generally unexplored, information of the soil microbial biology can be utilized to give pieces of information with regard to the feasible starting point of an unknown sample (Dawson and Mayes, 2015).

Human decomposition is an extremely complex chemical and biological process which includes the interaction of a various range of enzymes, fungi, bacteria, and protozoa, which separately function within a prime environmental condition (Carter et al., 2007). After death, the deficiency of oxygen initiates a chain of actions which results in cell necrosis and the inception of autolysis and then decomposition. Enzymes can be damaged, crystallized, or coagulated at high temperature. The best temperature for enzymes is 37°C with a range between 40 and 50°C. The theory of the "rule of ten" states that with each 10°C upsurge in temperature, the velocity of the biochemical reaction doubles or even triples (Gill-King, 1997). It is generally believed that acidity, water, temperature, and accessibility of oxygen can restrain the inner biological and chemical progression of decomposition (Vass et al., 2002; Vass, 2011). Most researchers believed that temperature and moisture are the most significant dependent variables for the rate of decomposition, while others proposed a number of other variables (Megyesi et al., 2005; Campobasso et al., 2001). It has been suggested that insect and scavenger activity burial, extent of clothing, burial depth, soil pH, body size, trauma, sun exposure, and rainfall have a vital impact on the degree of humanoid decay (Simmons et al., 2010; Weitzel, 2005).

Decomposition begins instantly after demise and is characterized by impulsive postmortem alterations. Soft tissue that has not been preserved is subject to the events of autolysis and putrefaction. Next to the putrefaction, there is liquefaction and disintegration, leaving skeletonized remains articulated by ligaments. Skeletonization continues until finally only the harder impervious tissues of bone, teeth and cartilage remain. These remains are then subjected to inorganic chemical weathering (Dent et al., 2004).

Soils have diverse textures, sand (2–0.02 mm), silt (0.02–0.002 mm), and clay (<0.002 mm), and chemical compounds change from one place to another. Twelve chief classes are defined in numerous combinations according to the size of constituents by the United States Department of Agriculture textural triangle (Brown, 2003). It is known that particle size of soil could interrupt process of decomposition. In this situation, decay can be entirely inhibited or diminished to the least, in coarse-textured soil due to diffusion of gasses through the soil matrix (Tibbett et al., 2004; Fiedler and Graw, 2003). Contrarily, pseudomorph might be produced in coarse-textured and high wet soil due to the rise of calcium, manganese, and phosphorus. On the other hand, fine-textured soil has lesser rate of air transmission and consequently, oxygen-CO_2 exchange might not be adequate to produce aerobic microbes, and these circumstances give the reason for the dominance of anaerobes which are less proficient decomposers (Carter, 2005; Anderson et al., 1979; Tumer et al., 2013). As a result of these, decay can be delayed due to lack of gas exchange; moreover, adipocere formation can be observed around the body or inner organs (Hopkins et al., 2000; Forbes et al., 2005).

3.4.1 Microbial soil profiling

As of not long ago, advancement in soil microbiological ecosystem was thwarted by the lack of approaches to characterize the huge variety of microbiological communities in soil. This is mainly due to the nonculturability of most microbial cells—for instance, in soil, the lot which can be cultured in the research facility has been assessed to be 0.3% of the whole number of cells witnessed microscopically—and furthermore because of the issues engaged with separating microorganisms from complex and variable networks (Rastogi and Sani, 2011).

3.4.1.1 Phenotypic analysis

The direct determination of cell segments from whole microbial assemblages declares one critical approach for quantitative evaluation of microbial abundance and organization. The most frequently utilized approach for this is phospholipid fatty acid analysis (PLFA). The PLFA method depends on the way that phospholipids exist in the membranes of every single living cell and that diverse subsets of the microbial assemblage contain distinctive

PLFAs, or if nothing else difference in their fatty acid composition (Tunlid and White, 1992). Besides, PLFA measures just the living portion of the microflora; phospholipids are not utilized as storage material, and when discharged upon cell decease they are utilized as substrate by living microorganisms and within minutes to hours are processed to diglyceride and $(PO_4)_3$ (White et al., 1979). The assessment of PLFA form of an ecological sample, therefore, can be seen as an incorporated estimation of every single living microorganism displayed in that sample, giving a quantitative measure of the distinctive microbiological communities inside diverse microbial communities. Nonetheless, it ought to be highlighted that since most PLFAs exist in various concentrations in a systematically extensive variety of microorganisms (Ratledge and Wilkinson, 1988), it cannot be utilized to quantify particular species or genera inside whole microbial groups; it can be utilized just to determine changes in the relative abundance of exceptionally expansive scientific classifications. The main role of fungi is to decay organic materials via the production of a vast array of extracellular enzymes. By virtue of their huge biomass they represent an important part of the biological community nutrient pool.

3.4.1.2 Genotypic analysis

Over the previous years, molecular biological practices in view of examination of the heterogeneity of RNA and DNA orders have reviewed microbiological taxonomy and significantly extended our insight into the microbial biology of soil. These methods have been made promising by our capacity to extract and purify DNA and RNA from soil. Genes giving evidence on the occurrence and variety of a particular physiological group or representing the phylogenetic relationship of detected microorganisms with different life forms have been utilized. Among the most extensively utilized phylogenetic genes are those of the RNA of the small subunit of the ribosome, the 16S rRNA and 18S rRNA genes of prokaryotic and eukaryotic microorganisms, respectively (Woese, 1987). The 16S rRNA gene and homologous 18S rRNA gene hold preserved areas, existing in all microorganisms, and variable sections that can be utilized to differentiate between sequences from diverse strains or species. The rRNA sequence database was first collected from well-categorized strains from culture assemblages (Larsen et al., 1993) and now denotes a dynamic and regularly expanding dataset against which new arrangements may be analyzed. The database incorporates sequences from refined strains and additionally from cloned gene sequences from hereditary material separated straightforwardly from ecological samples (Pace et al., 1986). The hereditary assorted variety of soil bacterial groups has been contemplated by an assortment of nucleic acid hybridization methods, including quality test examination and restriction enzyme examination.

Advancement of the polymerase chain response (PCR) has reformed investigations of microbial nature and prompted new techniques for the examination of bacterial group structure and assorted variety in soil samples. Since the mid-1980s, the dominant part of PCR-based investigations of microbial biology have centered on the examination of bacterial 16S rRNA. The 18S rRNA molecule of fungi has likewise been utilized as an objective in natural investigations (Kowalchuck et al., 1997; Anderson and Kohn, 1998). In addition to rRNA-based investigations, the recognition and examination of specific functional genes can like be utilized to deliver valuable data on the microbial biology of ecologically essential procedures.

3.5 Pedospheric samples maximization by utilization of GIS and RS

In addition to the soil samples characterization for various physicochemical and biological factors, it is also necessary that certain technological means are utilized for better comprehension of the various phenomena on large scale. On the global scale, the harmonization of the soil map for world has also been done around three decades ago. The soil map harmonization was done by the addition of new data from all world regions by production of the 30 arc resolution by the collective efforts of different institutions and scientific forums. The harmonized world map was made available on the online system and was provided with a viewer and was present in a geographic information system (GIS) compatible mode (Nachtergaele et al., 2010). However, in forensic soil examinations, whether it is the physicochemical/biological analysis or the integration of GIS and remote sensing (RS), the presence of the predefined set of procedures specialized for forensic domain is inexistent. In this regard, the chosen methodologies comprise, depending upon the case, the presence of people having expertise in scientific origin and instrumentation and of course there is also a factor of national preferences.

Soil analysis and evidence is used to complete the complete picture of any crime scene and is deemed to be the part of some investigative jigsaw puzzle. As predicted by previous studies, the present-day soil analysis for forensic purposes comprises the combination of traditional methods for inorganic components, e.g., mineral content with organic such as wax markers (Dawson and Hillier, 2010; Carvalho et al., 2013), and the investigations based on the microbial data (Concheri et al., 2011). For the integration of different software, e.g., GIS and RS, for the measurement of the variability in spatial domains of the soils of a particular place, the best known approach is the utilization of the sampling methods based on grid-based procedures also referred as "georeferenced" pedospheric sampling. This type of sampling yields the spatial variograms. Soil samples and field states in homogenous and heterogeneous forms will influence the grid size. The following section gives a brief account of the utilization of the software-based approaches for soil-based investigations which can be made use of in future forensic analysis.

3.5.1 Organochlorine pesticides migratory pattern

Studies have revealed the flux of organochlorine pesticides (OCPs) between air and soils, their levels and the pattern of the spatial distribution in Pakistan. Pakistan, being an agro economy, is marked by massive utilization of the agrochemicals and pesticides, thus the migration of pesticide residues and daughter products expresses a heavier burden on environment, particularly the lithospheric regions. Results for the OCP presence confirmed their origin from both the past and the present activities in the immediate soils or the use of these OCPs in close proximity (Alamdar et al., 2014).

3.5.2 RS-aided precision agriculture

As the GIS and RS can be utilized for the contaminant-level detect and presence, in the same way, they can also be used for betterment of the agricultural techniques. The forensic aspect of this involves the better detection of the evidence in case of different crimes in the lands that have been analyzed via GIS or RS. Since RS, being a tool, is very efficient in provisioning of the cheap, flexible, comparatively enhanced spatial resolution, thus greater benefit can be driven from RS in the accurate determination of the infield soil characteristics for precision agriculture. Studies revealed that successful RS-based quantification of soils for various parameters has been done. Such quantifications involve a large-scale applicability varying from the simulated conditions—driven laboratory experiments to the field-based studies ranging up to satellite hyperspectral imagery (Ge et al., 2011).

3.5.3 World soil organic carbon estimation

Global carbon cycle is predominantly influenced by the terrestrial ecosystems located on the higher altitudes. Despite such a significant place, the estimations for world soil organic carbon (SOC) are not inclusive of the updates in the permafrost-influenced soils. Furthermore, there is also no representation of the uniquely occurring pedological events. However, the quantification of the SOC stocks can be done by Northern Circumpolar Soil Carbon Database (NCSCD) in the permafrost region. The polygon-based NCSCD digital database has been made from the regional scale soil classification maps in harmonized form. In this database, the advancements have been done for spatial frameworks along with the potential applications (Hugelius et al., 2013).

3.6 Case study

Forensic comparisons have been made by analysis of soils for different elements in different regions of the world. A study was done based on the forensic investigations of garden soils in Pakistani and Turkish regions in the field conditions and was analyzed via inductively coupled plasma mass spectrometry. The pedospheric sample was done at different time intervals, i.e., 1, 5, 10, 20, 30, and 60 days. The undertaken work focused on the development of better procedure along with the derivation of error-free results. Pedospheric analysis of Pakistani and Turkish soils was done and statistically evaluated. The results were indicative of the presence of the different elements in the blood-stained garden soils. These soil samples obviously deviated from the negative control samples (forensic soil samples) and were not blood-stained. Results also favored the fact that the forensic soil investigations must be evaluated in the same specimen run, or if this is not possible then sequential runs must be assured for the accurate results from crime scene (Shahzad et al., 2016).

References

Abbasi, M.A., Butt, M.B., 2018. Soil fertility status of cultivated lands in different agro-ecological zones of Azad Jammu and Kashmir, Pakistan. Journal of Applied Agriculture and Biotechnology 2 (2), 21–27.

Ahmad, K.S., 2018a. Evaluating the adsorption potential of alachlor and its subsequent removal from soils via activated carbon. Soil and Sediment Contamination: International Journal 24 (4), 249–266.

Ahmad, K.S., 2018b. Sorption and Juglans regia-derived activated carbon-mediated removal of aniline-based herbicide Alachlor from contaminated soils. Environmental Earth Sciences 77 (12), 1–10, 437.

Ahmad, K.S., Jaffri, S.B., 2018a. Phytosynthetic Ag doped ZnO nanoparticles: semiconducting green remediators. Open Chemistry 16 (1), 556–570.

Ahmad, K., Jaffri, S.B., 2018b. Carpogenic ZnO nanoparticles: amplified nanophotocatalytic and antimicrobial action. IET Nanobiotechnology. https://doi.org/10.1049/iet-nbt.2018.5006.

Ajaz, M., Jabeen, N., Akhtar, S., Rasool, S.A., 2005. Chlorpyrifos resistant bacteria from pakistani soils: isolation, identification, resistance profile and growth kinetics. Pakistan Journal of Botany 37 (2), 381–388.

Alamdar, A., Syed, J.H., Malik, R.N., Katsoyiannis, A., Liu, J., Li, J., Jones, K.C., 2014. Organochlorine pesticides in surface soils from obsolete pesticide dumping ground in Hyderabad City, Pakistan: contamination levels and their potential for air–soil exchange. The Science of the Total Environment 470, 733–741.

Anderson, J.M., Heal, O.W., Swift, M.J., 1979. Decomposition in Terrestrial Ecosystems. Blackwell Scientific Publications.

Anderson, J.B., Kohn, L.M., 1998. Genotyping gene genealogies and genomics bring fungal population genetics above ground. Trends in Ecology & Evolution 13, 444–449.

Bardgett, R., 2005. The Biology of Soil: A Community and Ecosystem Approach. Oxford University Press.

Brown, R.B., 2003. Soil Texture. Fact Sheet of the Soil and Water Science Department, Florida Cooperative Extension Service. Institute of Food and Agricultural Sciences, University of Florida.

Campobasso, C.P., Di Vella, G., Introna, F., 2001. Factors affecting decomposition and Diptera colonization. Forensic Science International 120 (1–2), 18–27.

Carter, D.O., 2005. Forensic Taphonomy: Processes Associated with Cadaver Decomposition in Soil (Doctoral dissertation). James Cook University.

Carter, D.O., Yellowlees, D., Tibbett, M., 2007. Cadaver decomposition in terrestrial ecosystems. Naturwissenschaften 94 (1), 12–24.

Carvalho, A., Ribeiro, H., Mayes, R., Guedes, A., Abreu, I., Fernando Noronha, F., Dawson, L., 2013. Organic matter characterization of sediments in two river beaches from northern Portugal for forensic application. Forensic Science International 233, 403e415.

Concheri, G., Bertoldi, D., Polone, E., Otto, S., Larcher, R., Squartini, A., 2011. Chemical elemental distribution and soil DNA fingerprints provide the critical evidence in murder case investigation. PLoS One 6, e20222.

Dawson, L.A., Hillier, S., 2010. Measurement of soil characteristics for forensic applications. Surface and Interface Analysis 42 (5), 363–377.

Dawson, L.A., Mayes, R.W., 2015. Criminal and environmental soil forensics: soil as physical evidence in forensic investigations. In: Introduction to Environmental Forensics, third ed., pp. 457–486.

Dent, B.B., Forbes, S.L., Stuart, B.H., 2004. Review of human decomposition processes in soil. Environmental Geology 45 (4), 576–585.

Eilers, K.G., Debenport, S., Anderson, S., Fierer, N., 2012. Digging deeper to find unique microbial communities: the strong effect of depth on the structure of bacterial and archaeal communities in soil. Soil Biology and Biochemistry 50, 58–65.

Fiedler, S., Graw, M., 2003. Decomposition of buried corpses, with special reference to the formation of adipocere. Naturwissenschaften 90 (7), 291–300.

Forbes, S.L., Stuart, B.H., Dent, B.B., 2005. The effect of the burial environment on adipocere formation. Forensic Science International 154 (1), 24–34.

Gattinger, A., Palojärvi, A., Schloter, M., 2008. Soil microbial communities and related functions. In: Perspectives for Agroecosystem Management, pp. 279–292.

Ge, Y., Thomasson, J.A., Sui, R., 2011. Remote sensing of soil properties in precision agriculture: a review. Frontiers of Earth Science 5 (3), 229–238.

Gill-King, H., 1997. Chemical and ultrastructural aspects of decomposition. In: Forensic Taphonomy: The Postmortem Fate of Human Remains, pp. 93–108.

Hameed, S., Yasmin, S., Malik, K.A., Zafar, Y., Hafeez, F.Y., 2004. Rhizobium, Bradyrhizobium and Agrobacterium strains isolated from cultivated legumes. Biology and Fertility of Soils 39 (3), 179–185.

Harrison, R.B., Footen, P.W., Strahm, B.D., 2011. Deep soil horizons: contribution and importance to soil carbon pools and in assessing whole-ecosystem response to management and global change. Forest Science 57 (1), 67–76.

Hopkins, D.W., Wiltshire, P.E.J., Turner, B.D., 2000. Microbial characteristics of soils from graves: an investigation at the interface of soil microbiology and forensic science. Applied Soil Ecology 14 (3), 283–288.

Hugelius, G., Tarnocai, C., Broll, G., Canadell, J.G., Kuhry, P., Swanson, D.K., 2013. The Northern Circumpolar Soil Carbon Database: spatially distributed datasets of soil coverage and soil carbon storage in the northern permafrost regions. Earth System Science Data 5 (1), 3–13.

Ifthikhar, S., Ahmad, K.S., Jaffri, S.B., 2018. Mycodriven enhancement and inherent phytoremediation potential exploration of plants for lithospheric remediation. Sydowia 70, 141–153.

Ilyas, N., Bano, A., 2010. Azospirillum strains isolated from roots and rhizosphere soil of wheat (*Triticum aestivum* L.) grown under different soil moisture conditions. Biology and Fertility of Soils 46 (4), 393–406.

Iram, S., Ahmad, I., Stuben, D., 2009. Analysis of mines and contaminated agricultural soil samples for fungal diversity and tolerance to heavy metals. Pakistan Journal of Botany 41 (2), 885–895.

Iram, S., Parveen, K., Usman, J., Nasir, K., Akhtar, N., Arouj, S., Ahmad, I., 2012. Heavy metal tolerance of filamentous fungal strains isolated from soil irrigated with industrial wastewater. Biologija 58 (3).

Jaffri, S.B., Ahmad, K.S., 2017. Augmented photocatalytic, antibacterial and antifungal activity of prunosynthetic silver nanoparticles. Artificial Cells, Nanomedicine, and Biotechnology 1–11. https://doi.org/10.1080/21691401.2017.1414826.

Jaffri, S.B., Ahmad, K.S., 2018a. Prunus cerasifera Ehrh. fabricated ZnO nano falcates and its photocatalytic and dose dependent in vitro bio-activity. Open Chemistry 16 (1), 141–154.

Jaffri, S.B., Ahmad, K.S., 2018b. Phytofunctionalized silver nanoparticles: green biomaterial for biomedical and environmental applications. Reviews in Inorganic Chemistry 38 (3), 127–149.

Jaffri, S.B., Ahmad, K.S., 2018c. Neoteric environmental detoxification of organic pollutants and pathogenic microbes via green synthesized ZnO nanoparticles. Environmental Technology 1–42 (just-accepted).

Jaffri, S.B., Ahmad, K.S., 2018d. Foliar-mediated Ag: ZnO nanophotocatalysts: green synthesis, characterization, pollutants degradation, and in vitro biocidal activity. Green Processing and Synthesis. https://doi.org/10.1515/gps-2018-0058.

Khan, F., Hayat, Z., Ahmad, W., Ramzan, M., Shah, Z., Sharif, M., Hanif, M., 2013. Effect of slope position on physico-chemical properties of eroded soil. Soil & Environment 32 (1), 22–28.

Khan, M.A., Khan, S., Ding, X., Khan, A., Alam, M., 2018b. The effects of biochar and rice husk on adsorption and desorption of cadmium on to soils with different water conditions (upland and saturated). Chemosphere 193, 1120–1126.

Kheyrodin, H., 2014. Important of soil quality and soil agriculture indicators. Academia Journal of Agricultural Research 2 (11), 231–238.

Kowalchuck, G.A., Stephen, J.R., de Boer, W., Prosser, J.I., Embley, T.M., Woldendorp, J.W., 1997. Analysis of ammonia-oxidising bacteria of the -subdivision of the class Proteobacteria in coastal sand dunes by denaturing gradient gel electrophoresis and sequencing of PCR-amplified 16S ribosomal DNA fragments. Applied and Environmental Microbiology 63, 1489–1497.

Larsen, N., Olsen, G.J., Maidak, B.L., McCaughey, M.J., Overbeek, R., Woese, C.R., 1993. The ribosomal database project. Nucleic Acids Research 21, 3021–3023.

Megyesi, M.S., Nawrocki, S.P., Haskell, N.H., 2005. Using accumulated degree-days to estimate the postmortem interval from decomposed human remains. Journal of Forensic Sciences 50 (3), 1–9.

Mehdi, S.M., Sarfraz, M., Qureshi, M.A., Ilyas, M., Zaka, M.A., Qazi, M.A., Rafa, H.U., 2018. Site-specific phosphorus management with inorganic fertilizer and municipal solid waste compost application in salt affected soil. Pakistan Journal of Agricultural Sciences 55 (1), 103–110.

Moritsuka, N., Matsuoka, K., Katsura, K., Sano, S., Yanai, J., 2014. Soil color analysis for statistically estimating total carbon, total nitrogen and active iron contents in Japanese agricultural soils. Soil Science and Plant Nutrition 60 (4), 475–485.

Morris, M.W., Comeau, A., Ammons, J.T., Blair, R., Lu, C., Roberts, C., 2010. A comparion of three soil characterization methods on a soil formed in Sandy Glacial Outwash. In: Proceedings of the Annual International Conference on Soils, Sediments, Water and Energy, vol. 15, No. 1, p. 28.

Nachtergaele, F., van Velthuizen, H., Verelst, L., Batjes, N.H., Dijkshoorn, K., van Engelen, V.W.P., Montanarela, L., 2010. The harmonized world soil database. In: Proceedings of the 19th World Congress of Soil Science, Soil Solutions for a Changing World, Brisbane, Australia, 1-6 August 2010, pp. 34–37.
Pace, N.A., Stahl, D.A., Lane, D.J., Olson, G., 1986. The analysis of natural microbial populations by ribosomal RNA sequences. Advances in Microbial Ecology 21, 59–68.
Rastogi, G., Sani, R.K., 2011. Molecular techniques to assess microbial community structure, function, and dynamics in the environment. In: Microbes and Microbial Technology. Springer, New York, NY, pp. 29–57.
Ratledge, C., Wilkinson, S.G., 1988. Microbial Lipids, vol. 1. Academic Press, NY.
Schroder, P., 2008. Perspectives for Agroecosystem Management. Elsevier.
Sensabaugh, G.F., 2009. Microbial community profiling for the characterisation of soil evidence: forensic considerations. In: Criminal and Environmental Soil Forensics. Springer, Dordrecht, pp. 49–60.
Shahzad, M.S., Cengiz, S., Iqbal, J., Arif, A., Wattoo, J.I., 2016. Forensic soil analysis of gardens from Pakistan and Turkey for genetic analysis and elemental composition using ICP-MS. Forensic Research & Criminology International Journal 3 (2), 00092.
Sheik, C.S., Mitchell, T.W., Rizvi, F.Z., Rehman, Y., Faisal, M., Hasnain, S., Krumholz, L.R., 2012. Exposure of soil microbial communities to chromium and arsenic alters their diversity and structure. PLoS One 7 (6), e40059.
Siddiqui, F.I., Osman, S., 2012. Integrating geo-electrical and geotechnical data for soil characterization. International Journal of Applied Physics and Mathematics 2 (2), 104.
Siddiqui, F.I., Osman, S.B.A.B.S., 2013. Simple and multiple regression models for relationship between electrical resistivity and various soil properties for soil characterization. Environmental Earth Sciences 70 (1), 259–267.
Simmons, T., Adlam, R.E., Moffatt, C., 2010. Debugging decomposition data—comparative taphonomic studies and the influence of insects and carcass size on decomposition rate. Journal of Forensic Sciences 55 (1), 8–13.
Tibbett, M., Carter, D.O., Haslam, T., Major, R., Haslam, R., 2004. A laboratory incubation method for determining the rate of microbiological degradation of skeletal muscle tissue in soil. Journal of Forensic Sciences 49 (3). JFS2003247-6.
Tumer, A.R., Karacaoglu, E., Namli, A., Keten, A., Farasat, S., Akcan, R., Odabaşi, A.B., 2013. Effects of different types of soil on decomposition: an experimental study. Legal Medicine 15 (3), 149–156.
Tunlid, A., White, D.C., 1992. Biochemical analysis of biomass, community structure, nutritional status, and metabolic activity of microbial communities in soil. In: Stotzky, G., Bollag, J.M. (Eds.), Soil Biochemistry, vol. 7. Marcel Dekker, NY, pp. 229–262.
Vass, A.A., 2011. The elusive universal post-mortem interval formula. Forensic Science International 204 (1–3), 34–40.

Vass, A.A., Barshick, S.A., Sega, G., Caton, J., Skeen, J.T., Love, J.C., Synstelien, J.A., 2002. Decomposition chemistry of human remains: a new methodology for determining the postmortem interval. Journal of Forensic Sciences 47 (3), 542–553.

Vogel, H.J., Weller, U., Schlüter, S., 2010. Quantification of soil structure based on Minkowski functions. Computers & Geosciences 36 (10), 1236–1245.

Weitzel, M.A., 2005. A report of decomposition rates of a special burial type in Edmonton, Alberta from an experimental field study. Journal of Forensic Sciences 50 (3), 1–7.

White, D.C., Davis, W.M., Nickels, J.S., King, J.C., Bobbie, R.J., 1979. Determination of the sedimentary microbial biomass by extractable lipid phosphate. Oecologia 40, 51–62.

Woese, C.R., 1987. Bacterial evolution. Microbiological Reviews 51, 221–271.

Further reading

Khan, I.A., Arsalan, M.H., Ghazal, L., Siddiqui, M.F., Mehdi, M.R., Zia, I., Salam, I.U., 2018a. Satellite based assessment of soil moisture and associated factors for vegetation cover: a case study of Pakistan and adjoining regions. Pakistan Journal of Botany 50 (2), 699–709.

Soil Science of America. (December 1, 2014). Retrieved September 15, 2018, from: https://soilsmatter.wordpress.com/2015/12/01/what-is-soil-forensics/.

CHAPTER

Forensic evaluations of child abuse

4

Humaira Arif, MPhil, Anam Qaisar, MPhil, Shazia Iftikhar, PhD
Department of Environmental Sciences, Fatima Jinnah Women University, Rawalpindi, Pakistan

Chapter outline

Trends of Environmental Forensics in Pakistan. https://doi.org/10.1016/B978-0-12-819436-2.00004-2

Abstract

This chapter explains child abuse specifically in relation to forensics. At the start, it discusses various types of child abuse and its causes. Then it focuses on forensic aspects of child abuse by discussing the role of a child psychiatrist in this respect. Importance of chain of custody maintenance is also explained then. Forensic physical examination along with the patterns of skeletal fractures in child abuse is mentioned. At the end, how the documentation and reporting should be done of an alleged child abuse is explained. Moreover, laws related to child abuse in Pakistan are stated. Some case studies of child abuse in Pakistan are also highlighted.

Keywords: Child abuse; Forensics; Laws; Pakistan; Physical examination; Skeletal fractures.

Graphical Abstract

4.1 Introduction

Child abuse is an emerging social and health issue nowadays worldwide and in Pakistan as well. There is no pre-written book in Pakistan which addresses this issue of child abuse in relation to forensics especially. This chapter is compiled in a well-explaining manner. Child abuse, its types, sign and symptoms, and then its causes are addressed first. Then forensics role in child abuse is explained. What are the steps to be taken when a child abuse is suspected are also listed. Laws related to child abuse in Pakistan are also explained. Along with all these important aspects, different case studies of child abuse are also mentioned along with the most recent ratio and number of child abuse cases in Pakistan. So, this chapter will provide the readers comprehensive information related to child abuse in relation to forensics, laws and regulations, and case studies and updated data of child abuse in Pakistan.

4.2 Child abuse

The child abuse and negligence are the practices in which a child is ill-treated. The child is abused or neglected both directly and indirectly. By direct means the harm or injury may be imposed on a child forcefully, and by indirect means there is failure to protect child from harm. Children may possibly be abused or maltreated either within a family or in an institute or within a public locality, and these abusers may be the child's familiar person or may be a stranger. The child is abused either individually or by a group of people, and both adults and children are involved in it.

Diagnosing child manipulation is not an easy task. One cannot decide on his own whether the child has been abused or not or the child is at a risk of abuse. It is the responsibility of the child's protection organization to help the child under the procedures set by them. These organizations can provide better protection to child. The succeeding knowledge may help to get aware of the types of abuse and signs and symptoms of abuse.

4.3 Different types of child abuse and their Signs and symptoms

The child abuse is classified into five classes and these are discussed in the following sections.

4.3.1 Physical abuse

In physical abuse, the child may be hit, shivered, thrown, poisoned, burnt or scorched, drowned, suffocated, or else the child may be physically harmed or failed to be protected from the harm. This kind of harm may also be triggered even if a parent or guardian produces the conditions of inducing the harm intentionally in a child.

4.3.1.1 Signs and symptoms of physical abuse

Signs and symptoms of physical abuse include bruises on arm and thigh, the marks of cigarette burn, marks of human bite, cracked bones, blisters and several injuries with a defined edge. Variations in child's attitude also point toward the physical abuse and may include the distress, violent behavior or temper bursts, flinching as soon as touched, unwillingness to get dressed up, hopelessness, reserved activities, and missing from family gatherings.

4.3.2 Emotional abuse

In this kind of emotional abuse, the child's emotions and emotional development are adversely affected, and the child is maltreated emotionally. This kind of abuse is done to make the children feel that they are valueless or they are not loved by anyone, poor, or they can only be loved and appreciated if they fulfill the requirements or needs of somebody. The children are intentionally imposed to involve in these kinds of activities which affect their development proficiency. The child is over protected and his/her access to education is narrowed down and the child is prevented to participate in social or communal dealings. The children are repeatedly scared and exploited emotionally.

4.3.2.1 Signs and symptoms of emotional abuse

It is very challenging to measure the symptoms of emotional abuse, as there are no prominent external physical marks. There might be a failure in development and growth if he/she is away from parents' care, love, and attention, although this will generally only be obvious if the child gains the weight. The children may be emotionally hurt if they are not permitted to play with their age fellows and mix with the gatherings. Fluctuations in behavior which point out emotional abuse may include phobic actions, e.g., high temper, twisting the hair, disturbing attitude, not capable to perform in extracurricular activities, losses self-confidence, communication syndromes, fear from parents, and failure in emotional development.

4.3.3 Sexual abuse

In this type of abuse the child or may be young ones are forced to involve/indulge in sexual deeds or actions. Prostitutions are also included in this kind of abuse, and sometimes the children are unconscious of these activities.

The direct activities may involve the kissing act by touching and patting the sex organs and private areas or having the intercourse or may be the oral sex. And the indirect means may be by providing the children inappropriate material related to sexual activities or they may be forced to watch the sexual ways.

4.3.3.1 Signs and symptoms of sexual abuse

In sexual abuse cases there are marks on external body, but the child's behavior is the most alarming one. In most cases children want to speak, but they are not given attention properly. The bodily signs of sexual abuse include discomfort or irritation in the genital region, marks or exploited genital cavity, sexually transmitted infections, vaginal infections, stomachaches, distress while walking or sitting, and the worse symptom is the pregnancy. Alterations in behavior include aggression, fear, nightmares, sexual awareness beyond their developmental stages, sexual pictures, bed-wetting, overeating, lack of confidence, suicide attempts, drug abuse, and a possible secret source of cash.

4.3.4 Neglect

When a child's physical or psychological requirements are not fulfilled, then there is a serious damage to a child's healthiness or growth. Neglect may ensue during the time of pregnancy due to maternal negligence. Neglect may ensue when a parent fails to afford suitable nourishment, clothing, and housing for a child and also fails to shelter the child from both physical and emotional harm; fails to guarantee suitable caretakers; fails to make access to adequate medical healthcare. If a child's emotional requirements are not responded appropriately, then it also comes under neglect.

4.3.4.1 Signs and symptoms of neglect

It is difficult to recognize this kind of abuse, but the side effects of the neglect is the most damaging one and last long. The signs and symptoms include persistent hunger, stealing food from others, not dressed up in an appropriate way, unclean clothes, conditions of weight loss, etc. Other signs and symptoms may also include sleepy or drowsy conditions, not getting medical care, having less number of friends, and mostly complaining of left unaccompanied.

4.3.5 Bullying

It is well defined as intentionally unkind activities, frequently repeated with time, and making it hard for bullied to secure themselves. It has various types, but there are three leading forms: physical bullying (e.g., beating, jerking, robbery), verbal bullying (e.g., discriminatory comments, pressures, title giving), and emotional bullying (e.g., separating the child from social activities). The harm caused by bullying cannot be taken too lightly, it may result in significant problems to the child's health and also affects the developmental stages of child.

4.3.5.1 Signs and symptoms of bullying

Bullying can cause hopelessness, lack of self-confidence, nervousness, poor educational attainment, loneliness, and vulnerable or attempted suicide. Signs and symptoms include wounds and bumps, torn dresses, inquiring for means of replacing stolen assets, losing money, not having respectable friends, moody and poor tempered, avoiding to leave home, violent behavior with siblings, nightmares, uneasiness, becoming unobtrusive and reserved.

4.4 Reasons of child abuse

Children nowadays may experience such maltreatments due to numerous reasons. In a community, the religious and traditional practices are also involved in child abuse. Illiterate people are not aware of negative effects of these practices. It can be best explained by the example of early child marriages which affects the child's physical and emotional development. Although people think that early child marriages may bring a gratitude for their family, but this is not the case, the child's development suffers a lot and emotions are directly harmed. The Holy book also promotes the concept of love and friendliness for the children. There are a number of reasons for child abuse.

4.4.1 Poverty

This is a main reason of child abuse. Some parents are in such a poverty state that they even cannot provide the basic needs and requirements to their children, so they allow their children to be exposed to such abusive practices to earn money for their family. In such a state the children are exposed to strangers where they are offensively misused or mistreated. Their employers misuse them because they know that these children are not protected by anyone, so they get benefit to fulfill their desires.

4.4.2 Vulnerability

Several children are targets or victims of child abuse because they cannot defend themselves and are prone to vulnerability. The abuser gets the benefit of it that the children cannot defend themselves and by threatening them they do not allow the child to speak for their harm and damage their emotions. Many children are vulnerable to abuse because of their disability or he/she may be found living in streets or in many cases when they are not living with their actual parents.

4.4.3 Wickedness

In several cases, the abuser is criminal and terrific and his/her desires cannot be fulfilled in any further approach and these are mentally sick people. Most of the abusers abuse the children for their own sickening enjoyment. By abusing children, they fulfill their sense of happiness.

4.4.4 "I don't care" attitude

This kind of attitude is shown by the people who are the eye witnesses of this abusive activity. Because they do not want to indulge in the authorities' procedures, they close the eyes and act like they have not seen anything (Child Welfare Information Gateway, 2013).

4.5 Child abuse in relation to forensics

Forensics plays a very important and efficient role in child abuse investigation. Based on the evidences collected by forensic experts and reports which are made on the basis of these evidences, an alleged abuse is proved or disapproved and the criminal is sentenced for this shameful and heart-throbbing act which is necessary to prevent the child abuse.

4.6 Forensic role of child psychiatrist

Although psychiatrists know their role in clinical settings, there are chances that some do not save this experience of forensic matters. The purpose of explaining forensic role of a child psychiatrist is to make them more aware of their work for child abuse cases. Mostly, the child psychiatrist's role in forensic cases depends on the boundaries made during the early contact with the engaging consultant. This must be done in the first meeting or phone contact for an efficient forensic assessment. In criminal cases, the state's attorney, defense attorney, or the court itself may seek to hire the child psychiatrist. But, in civil cases, a person, often an attorney, appointed by the court, private attorney, law enforcement investigators, Child Protective Services, may request his or her services to represent the "best interests of the child". Sexually abused children are mostly asked for forensic assessment as these cases mostly depend upon the statements and behaviors of the child. It is the most crucial duty of the child psychiatrist to interview the child and is a fundamental step for the inquiry of the case. Here, different interview techniques will be reviewed. Most importantly, the steps of the interview which will be very useful intended for forensic assessors include the following:

1. Constructing a relationship with the child.
2. Requesting him/her to explain specific events not linked to the suspected abuse.
3. Try to know ability of the child to separate truth from lie and then compel him/her to speak the truth only.
4. Present the topic of discussion with common questions.
5. Encourage his/her to explain the incident since the start without disruption.
6. Ask particular questions in a nonsuggestive manner.
7. Conclude the interview with some important questions about unrelated topics to test his/her liability to suggestion.

Child psychiatrists perform significant roles in the judiciary. They assist in giving the impartial information to the judge and panel intended for legal decisions in child abuse cases. A forensic psychiatrist must know the moral values to be followed. Finally, they must keep in mind that the role of the child psychiatrist in forensic matters is dynamic which is formed by the continuously developing behavior of the society, law, and the medical profession in relation to the child abuse (Leavitt and Armitage, 2002).

4.7 Chain of custody in child abuse cases and its importance

Maintenance of chain of custody consists of protecting the integrity of the evidence by means of documentation of its custody from the time it was gathered up till it is presented in the court. For making evidence safe in specified place in the operating room and for its gathering, classification, and conservation, perioperative nurses are in charge. It is very important to follow these stages to keep the chain of custody. For the maintenance of chain of custody, it is important to consider evidence assortment kit and samples that are present in it from the time of gathering till the time it is presented in the court. After gathering the specimens, no one, whether he is parent, support person, victim, CPS caseworker, or guardian, should be permitted to hold or transport it in any condition. There should be a proper method for preservation of chain of custody at every healthcare facility. Physical evidence should be transferred to crime laboratory facility for analysis by a law enforcement official, and its documentation should be done and record must be saved. Hair, clothing, debris from patients' bodies are some examples of forensic evidences.

Samples must not be contaminated during their collection; this is essential for a transparent way of investigation. Gloves should be used to hold the evidences. Vital trace evidences are usually present on the clothing, so it should be detached, preserving the holes in it. Clothing should not be shaken as it may consist of essential forensic evidences, e.g., fibers, DNA, hair, or blood. To avoid cross-contamination, every sample should be packed in a distinct paper bag having a white sheet of paper located on the stains, and especially clothing should not be stored in a plastic bag to avoid cross-contamination. Evidence tape should be used to seal the bags, and patient's name, date, time, and signature of the individual answerable for gathering and preserving the samples/evidences should also be written on every sample especially on the wet ones to make crime laboratory personnel aware of the fact that a wet sample is present. Wet samples should be picked up instantly. Wet evidences should not be placed in a drop box for storage.

There should be a refrigerator in a locked drop box for the placement of samples. It is the finest practice. This facility should be situated in an area where there is restricted public entry and entrance is allowed only to the specified persons only. These are very essential points for the maintenance of chain of custody.

4.8 Forensic physical examination

Characteristically, a child welfare authority conducts the forensic interview of the child in cases of suspected abuse. Ideally, this would occur before the medical assessment. A full medical forensic examination includes taking a patient history and conducting a head-to-toe assessment for trauma, including a detailed genital examination, to evaluate forensic evidence.

Physical examination of a child consists of different steps like documenting the growth parameters together with the weight and height as well as head perimeter for the children of less than 48 months of age. A head-to-toe examination must be carried out exposing all parts of skin and imaging of the buttocks and genitals.

If possible, documentation of skin results with suitable photos is supportive and enables peer or external review. A color bar and measurement scale should be included in the snapped field when possible.

A strong complement strategy should be logged, consisting of suggestions for action, supplementary medical testing, timing and necessity for reconsideration, and, if possible, whether a report was or will be made to a child welfare authority. A review of one's documentation for precision and wholeness is suggested (Ornstein, 2013).

4.9 Patterns of skeletal fractures in child abuse

There is a strong chance of fracture due to abuse depending upon the age and developmental stage of the child (Bishop et al., 2007).

4.9.1 Rib fractures

Children who had been abused had greater number of rib fractures than those who had not been abused. These fractures could be unilateral or bilateral. In abused children anterior fractures are common, but in nonabused ones, there is greater ratio of lateral fractures.

4.9.2 Skull injury

These fractures are mostly stated after nonabusive injuries than the abusive ones. Parietal site is the most common site for fracture in both groups. And linear fracture is the most common type of fracture. Bilateral multiple fractures have greater ratio in abused children. For complex, depressed, and diastatic fractures, there is not much difference between abusive and nonabusive children. For the children of age less than 3 years, there is a greater ratio of diastatic and multiple fractures in abused children.

4.9.3 Fractures of lower limbs

Midshaft of femur is the place having the highest ratio of fractures in both abused and nonabused group. There are greater chances of spiral fracture;

a type of femoral fracture in abused children of age less than 15 months. Above this age, there is no noteworthy difference of such fractures in both groups. Another type of fracture called metaphyseal fracture is more common in abused than in nonabused children (Kemp et al., 2008). A study by Caffey (1946) showed that 96% of fibular/tibial fractures are the consequence of abuse.

4.10 Documentation and reporting

Following points are important and must be followed for documentation as well as reporting of a crime scene (child abuse).

- Record what the child said in their own words and whether the child is answering spontaneously or not.
- Interview the parent(s) independently and record their description.
- Record what occurred, when, where, and how—any witnesses?
- Who lives with the child/takes care of the child?
- Note history of past injuries, hospitalizations, and medical conditions which might mimic abuse pattern.
- Note the physical and emotional state of the child when disclosing.
- Note cleanliness, condition, and suitability of clothing.
- Perform a complete physical examination with detailed documentation of any suspicious areas.
- If sexual abuse is suspected, do not perform a genital examination except for superficial visual inspection, as it may negate subsequent forensic examination evidence collection.
- Pictures can be very helpful.
- Document reasoning why injury appears to be nonaccidental.
- Document who made the verbal and written report, when and to whom, and the name and/or badge number of Child Protective Services or police involved.
- Document recommendations made for other services and follow-up appointment, such as counseling.
- Document information given to nonabusive parent, including medical care instructions.
- Document nature of the child (Stanford Medicine, 2018).

The form and place of wounds should be documented properly to reserve the particulars of the wounds. In an ideal situation, there is enough time to take photos to document any injuries (e.g., bruises, lacerations). There

FIGURE 4.1

Steps involved in investigation of an alleged child abuse.

must be a ruler in photos or any article that shows scale, e.g., a quarter. Photos must be labeled by patient information and then placed in a sealed cover and put in the locked drop box (Evans and Stagner, 2003).

The clinician should note who has brought the child for examination. It is significant for the clinician to document both the subjective (i.e., parental worry about possible sexual abuse) and objective information (i.e., presence or absence of vaginal discharge on examination). Clinicians are requested to be aware of the rules and regulations about the child protection and their respective practice area (Omstein, 2013) (Fig. 4.1).

4.11 Laws related to child abuse in Pakistan

1. Punishment for physical abuse: Whosoever is responsible for physical abuse of child, through any form of command, under any condition or reason, imprisonment shall be imposed which may possibly extend to more than 6 months or may be punished to pay a fine of 50,000 rupees, or in some conditions fine and imprisonment both may also be imposed.

2. Cheating on a child: Whoever, by disputes, or by symbols or else, stimulates, attempts to stimulate, cheatings or permits a child to participate in any actions which may be injurious to the body, health, spiritual, expressive, financial, and societal welfare of a child will be imposed imprisonment, which might extend up to 3 years or with fine of more than 100,000 rupees or may be imposed with both punishments.
3. Penalty for act of violence against a child: Whoever is responsible for violence will be imposed imprisonment which might extend up to 3 years or with fine of more than 100,000 rupees or may be with both. And if more than two people are responsible for this violence then each will be imposed incarceration up to 10 years and each will be fined with more than Rs. 100, 000.
4. Penalty for injurious and offensive practices: Whoever is involved in the crime will be imprisoned up to 3 years and may be fined up to 100,000 rupees or may be imposed with both punishments.
5. Penalty for trading body parts of a teenager: Whoever deals in trading of body organs will be sentenced to death or imprisoned for whole life and fined up to a million rupees.
6. Illegal custody: Whoever takes illegal custody of a child will be imprisoned up to 2 years or fined up to 50,000 rupees or may be imposed with both punishments.
7. Being cruel with a child: Whosoever not actually a true parent and has the charge to control a child and intentionally stabs, abuses, mistreats, or exposes the child to physical abuse; or fails in providing sufficient food, garments, or medicinal aid; or hurts the child both mentally and physically will be given punishment with imprisonment for 3 years and fined with 1000 rupees.
8. Engaging child in begging act: Whosoever hires any child for begging will be imprisoned for 3 years and will be fined for 50,000 rupees.
9. Giving hallucinogenic liquid or tranquillizers to a child: Whosoever provides or involved in giving intoxicants or narcotics to a child, except the medical specialist, will be imprisoned up to 4 years and will be fined up to 50,000 rupees.
10. Punishment for child pornography: Whosoever is responsible for child pornography will be imprisoned with 3 years and can prolong up to 7 years and will be also legally responsible to pay a fine more than 200,000 rupees which can extend to 500,000 rupees.
11. Involvement in sexual behavior: Whosoever involves a child in any kind of sexual activities or provides any kind of clear sexual material,

manuscript, picture, film or tries to do the abovementioned actions, will be imprisoned for many years or fined for a rupee of 100,000.

12. Child trading: Whosoever is involved in child trading in Pakistan will be imprisoned for whole life or for more than 14 years and will be responsible to pay a fine of 500,000 rupees which may extend up to 1,000,000.
13. Sexual misuse: Whosoever involves in sexual misuse of a child will be imprisoned up to 14 years, may be not less than 7 years, and will be responsible to pay a fine of 1,000,000 rupees.

4.12 Case studies of child abuse in Pakistan

There is an alarming ratio of child abuse in Pakistan. A research says that there is 10% increase in child abuse in Pakistan (*The Express Tribune*, Sehrish Wasif, 2017). Another study says that roughly 11 children are abused every day in Pakistan. A report made by the Sahil, a nongovernment organization working against child abuse, reports that a total of 4139 cases were reported in 2016, an increase from 3768 in 2015. The report says that 1729 boys and 2410 girls were victims of sexual exploitation in 2016. Almost 45% of the victims were between the ages of 6 and 15 years (Samawiya Javed, The Nation, 2017). Results of another annual report by Sahil organization named "Cruel Numbers 2017" showed that in the year 2017, a total of 3445 child abuse cases were registered with an average of 9 cases/day. Results also showed that 60% victims were girls and 40% were boys. The report is compiled on the basis of child abuse cases reported in 91 newspapers from all four provinces including Islamabad Capital Territory, Azad Jammu and Kashmir, Gilgit-Baltistan, and Federally Administered Tribal Areas (Myra Imran, The News International, 2018).

4.12.1 Zakir-Sahil case study

On June 16, 2010, in the police station of Fateh Jang, Attock, a case was registered. On May 25, 2010, the suspects named Kamran and Amir approached to the family of Zakir and requested his father to allow Zakir to work with them in their poultry farm. His father gave permission to Zakir because it was a good chance to generate a handsome income for family. Zakir's age was just 16 years. On June 9, 2010, Zakir was sodomized by both Kamran and Amir. On June 15, 2010, Zakir successfully ran away from that farm and came back to home in totally scared state and stated to father that Kamran and Amir sodomized him repetitively for 3 days. He also told that they beat him with a pipe of water and also with an iron rod. His father instantaneously filed for FIR against both suspects. A complete enquiry and hearing was headed in the high court. The decision was given away on October 9, 2010, and it was stated that both the suspects were found to be guilty. Both the suspects were imprisoned for 3 years with a fine of 10,000 rupees for each one (Sahil Legal Case Studies, 2014).

4.12.2 Anisa-Sahil case study

On October 16, 2009, a case was listed at the police station of Faiz Ganj. The FIR was filed by Anisa, the victim, who was a student of matric class at Government High School, Bhangu. On October 10, 2009, she went to school with her uncle. One of the partners of crime named Sana, at 9'o clock in the morning requested her to get along with her to bring her homework copies. They went to another crime scene partner's home, there Iftikhar was standing outside his house and Anisa was left in a room. Anisa's teachers named Mustafa, Ilyas, and Shuja were also present inside the room. They immediately closed the room's door. One of the suspect Shuja told her that as she refused to have an affair with him, this was the time of revenge and now they will not spare her.

She stated that they hold her arms and tied her mouth with her doppata. Then the suspect Shuja committed sexual abuse with her by removing her clothes too. In the meanwhile, other two witnesses named Farrukh and Abdullah, came Iftikhar's home in search of Anisa. They inquired for Anisa from Iftikhar and Sana and after getting no response from them they started searching for Anisa inside the house of Iftikhar. Somehow, they succeeded in opening the window of the room and they were shocked to saw the teachers committing the zina with her. The suspects fled from the crime scene and the witnesses managed to untie the mouth and arms of Anisa. Then Anisa narrated the whole incident to the witnesses. The witnesses then brought her in front of the head of the school. There in the office her father and other blood relatives were called. Then, Anisa narrated the whole horrific incident in front of her father. The suspects tried to make concessions, but her father did not agree upon their commitments and filed the FIR after 6 days of the incident.

After getting free legal aid from Sahil, the lawyer progressed the case in the high court. After thorough hearing from the Session Judge on December 4, 2010, all of the three suspects Mustafa Shuja, and Ilyas with their crime partners, Iftikhar and Sana, were imprisoned for 25 years with a fine of 25,000 rupees on each member (Sahil Legal Case Studies, 2014).

4.12.3 Zainab Ansari rape and murder case

A 7-year-old girl Zainab was kidnapped on January 4, 2018. CCTV footage emerged which showed the child, who was on her way to a religious studies class, being led away by a man on the evening of January 4. Her body was found in a rubbish dump on January 9, and it was found that she has been died for 2 days. The man has been given four death sentences for rape and murder of a 7-year-old girl in Pakistan last month. Zainab Ansari was abducted, sexually assaulted, and killed in Kasur, a city in the country's Punjab region. Imran Ali, 24, has also been linked to murders and sexual assaults of other girls in the region by police. Her parents were in Saudi Arabia to perform Umrah. The forensic scientist also confirmed that the suspect's DNA samples were matched with the samples taken from the body of the victim (Lydia Smith, The Independent, 2018).

References

Bishop, N., Sprigg, A., Dalton, A., 2007. Unexplained fractures in infancy: looking for fragile bones. Archives of Disease in Childhood 92 (3), 251–256.

Caffey, J., 1946. Multiple fracture in long bones of infants suffering from chronic subdural hematoma. American Journal of Roentgenology 56, 163–173.

Child Welfare Information Gateway, 2013. What Is Child Abuse and Neglect? Recognizing the Signs and Symptoms. U.S. Department of Health and Human Services, Children's Bureau, Washington, DC.

Evans, M.M., Stagner, P.A., 2003. Maintaining the chain of custody evidence handling in forensic cases. AORN Journal 78 (4), 563–569.

Kemp, A.M., Dunstan, F., Harrison, S., Morris, S., Mann, M., Rolfe, K., Maguire, S., 2008. Patterns of skeletal fractures in child abuse: systematic review. BMJ 337, 1518.

Leavitt, W.T., Armitage, D.T., 2002. The forensic role of the child psychiatrist in child abuse and neglect cases. Child and Adolescent Psychiatric Clinics of North America 11 (4), 767–779.

Lydia Smith, The Independent, 2018. Retrieved from: https://www.independent.co.uk/news/world/asia/zainab-ansari-rape-murder-man-imran-ali-death-sentence-pakistan-lahore-kasur-a8215151.html.

Myra Imran, The News International, 2018. Retrieved from: https://www.thenews.com.pk/print/300651-report-on-child-sexual-abuse-launched-3-445-children-abused-in-pakistan-last-year.

Ornstein, A.E., 2013. An approach to child maltreatment documentation and participation in the court system. Paediatrics and Child Health 18 (8), 44–47.

Sahil Legal Case Studies, 2014. Retrieved from: www.sahil.org.

Samawiya Javed, The Nation, 2017. Retrieved from: https://nation.com.pk/07-Apr-2017/child-abuse-in-pakistan.
Stanford Medicine, 2018. Retrieved from http://childabuse.stanford.edu/reporting.html.
The Express Tribune, Sehrish Wasif, 2017. Retrieved from: https://tribune.com.pk/story/1363150/rise-10-increase-child-abuse-cases-pakistan.

CHAPTER

Forensic materials analysis and engineering

5

Shania Sharif, PhD Scholar, Sara Majid, PhD Scholar, Shazia Iftikhar, PhD, Khuram Shahzad Ahmad, PhD

Department of Environmental Sciences, Fatima Jinnah Women University, Rawalpindi, Pakistan

Chapter outline

Abstract

Forensic material investigation covers the scientific and engineering doctrines to solve a crime. Engineering evaluation accomplishes the source of the failure to help assign responsibility for the damages. Most failures involve materials damage of some sort. The special skill required for forensic engineering is the ability of the analyst to break down the technical issues for interpretation by nontechnical people, such as attorneys, agents, judges, and juries. The main principle of forensic investigations is presence of actual trace evidence of a criminal or accidental activity which makes a record of itself. This chapter covers the analytic techniques, chemical markers, and methods involved in forensic material analysis.

Keywords: Engineering; Failure; Forensic; Interpretation; Investigation; Techniques.

Trends of Environmental Forensics in Pakistan. https://doi.org/10.1016/B978-0-12-819436-2.00005-4

Graphical Abstract

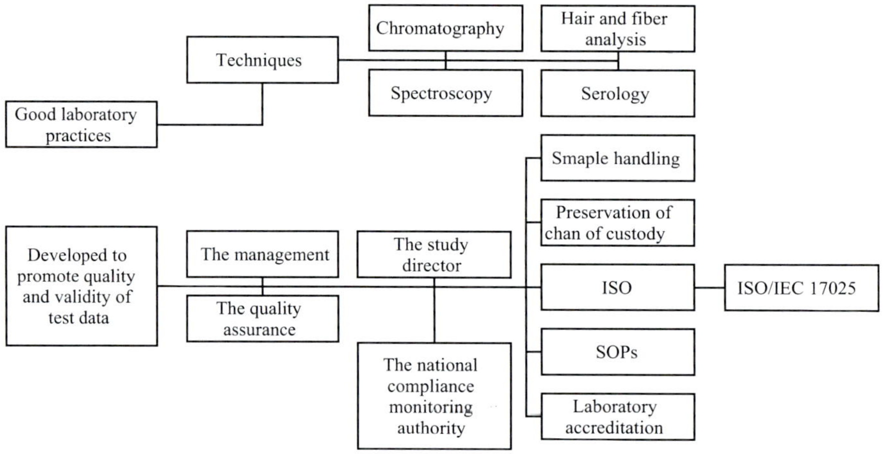

5.1 Introduction

Forensic material analysis is the application of scientific and engineering principles to settle disputes between two or more parties. These disputes typically arise when a product or structure fails to perform as expected and causes a loss to one of the parties. Forensic material analysis is practiced for insurance companies evaluating claims for payment or subrogation and for attorneys pleading or defending product liability law suits. Engineering evaluation determines the cause or causes of the failure to help assign liability for the damages. Most failures involve materials damage of some sort. Therefore, materials engineering is often an integral aspect of forensic engineering investigations. Evaluation of the materials behavior using standard failure analysis practices typically identifies the failure mode, and this information, on its own or in combination with other engineering evaluations, leads to determination of the root cause and liability. The special skill required for forensic engineering is the ability of the analyst to break down the technical issues for interpretation by nontechnical people, such as attorneys, agents, judges, and juries (Brooks and Choudhury, 2002).

5.2 Classification of forensic materials

Materials come under three basic categories:

- Metals
- Ceramics
- Polymers

A mixture of these fundamental types forms a composite. Two classification schemes are discussed in the following sections.

5.2.1 Type 1 classification

Engineering materials can be classified into six broad families as shown in Fig. 5.1.

FIGURE 5.1

Classification of materials.

- Metals
- Polymers
- Elastomers
- Ceramics
- Glasses
- Hybrid composite materials

5.2.2 Type 2 classification

In this type of classification, engineering materials can be classified into two categories, metals and nonmetals, as shown in Fig. 5.2. Nonmetals are classified into organic and inorganic. Metals can be classified into two categories: ferrous and nonferrous metals. Ferrous metals contain iron in them. Pure iron has limited use, but when alloyed with carbon it has a great commercial value. Some of the common alloys of iron are steel and cast iron which contains different percentages of carbon in it. Steel contains 0.02%–2.11% of carbon and iron, manganese, chromium, nickel, and molybdenum in it. Cast iron contains 2%–4% of carbon in it and the rest are iron and silicon. Nonferrous metals contain other metallic elements other than iron in it. They include aluminum, copper, gold, etc.

Ceramics are the compounds which contain a metallic and a nonmetallic part. The nonmetals can be oxygen, nitrogen, and carbon. Examples of ceramics include carbides, clay, silica, alumina, etc. Polymers are compounds

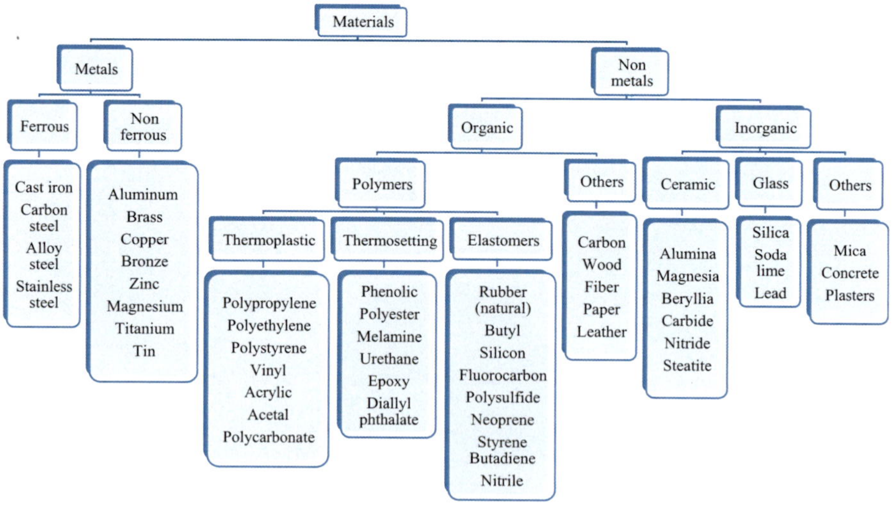

FIGURE 5.2

Classification of materials into groups.

which consist of repeating units in them called as "mers." Mers share electrons to form very large molecules—usually of carbon and some other elements like oxygen, hydrogen, nitrogen, chlorine, etc. Polymers are further classified into thermosetting, thermoplastics, and elastomers. Some of the common polymers are polythene, PVC, etc. Composites consist of two or more phases of materials. The phases are processed separately and then bonded together to achieve properties superior to the constituents. Some of the materials used in the phases are wood, fiber, etc., which are a homogenous mass bonded together with epoxy. Some of the common applications of composites are aircraft, tennis rackets, car bodies, etc.

5.3 Methods and principles for detection of trace evidences

The failure analysis of many different products involves the use of the following tools and techniques (Table 5.1).

Table 5.1 Techniques used in material analysis.

Techniques	Types
Microscopy	• Optical microscope • Scanning acoustic microscope (SAM) • Atomic force microscope (AFM) • Stereomicroscope • Photoemission electron microscope (PEM) • X-ray microscope • Infra-red microscope • Scanning SQUID microscope • USB microscope (Chalmers et al., 2012)
Sample preparation	• Jet etcher • Plasma etcher • Metallography • Back-side thinning tools • Mechanical back-side thinning • Laser chemical back-side etching (Demeestere et al., 2007)
Spectroscopic analysis	• Transmission line pulse spectroscopy (TLPS) • Auger electron spectroscopy • Deep-level transient spectroscopy (DLTS) • Fourier transform-infrared spectroscopy (FTIR) (Banas et al., 2010) • Raman spectroscopy (Fedicket al., 2017)

Continued

Table 5.1 Techniques used in material analysis.—*cont'd*

Techniques	Types
Device modification	• Focused ion beam etching (FIB) (Weiszburg et al., 2017)
Surface analysis	• Dye penetrant inspection • Other surface analysis tools (Skoog et al., 2017)
NMR analysis	• J-spectroscopy, exchange spectroscopy (EXSY) • Nuclear overhauser effect spectroscopy (NOESY) • Total correlation spectroscopy (TOCSY) • Heteronuclear correlation experiments such as HSQC, HMQC, and HMBC (Skoog et al., 2017)
Chromatography	• Column chromatography • Ion-exchange chromatography • Gel-permeation (molecular sieve) chromatography • Affinity chromatography • Paper chromatography • Thin-layer chromatography • Gas chromatography • Dye-ligand chromatography (Dunlop et al., 2017)
Thermal analysis	• Dielectric thermal analysis (DEA) • Differential thermal analysis (DTA) • Differential scanning calorimetry (DSC) • Dilatometry (DIL) • Dynamic mechanical analysis (DMA or DMTA) • Evolved gas analysis (EGA) • Laser flash analysis (LFA) • Thermogravimetric analysis (TGA) • Thermomechanical analysis (TMA) • Thermo-optical analysis (TOA) • Derivatography (Clayton, 2017)
Electron microscopy	• Scanning electron microscope (SEM) • Electron beam–induced current (EBIC) in SEM • Charge-induced voltage alteration (CIVA) in SEM • Voltage contrast in SEM • Electron backscatter diffraction (EBSD) in SEM • Energy-dispersive X-ray spectroscopy (EDS) in SEM • Transmission electron microscope (TEM) • Computer-controlled scanning electron microscope (CCSEM) (Ellingham et al., 2018)
Laser signal injection microscopy (LSIM)	• Photo carrier stimulation • Static • Optical beam induced current (OBIC) • Light-induced voltage alteration (LIVA) • Dynamic • Laser-assisted device alteration (LADA)

Table 5.1 Techniques used in material analysis.—*cont'd*

Techniques	Types
	• Thermal laser stimulation (TLS) • Static • Optical beam–induced resistance change (OBIRCH) • Thermally induced voltage alteration (TIVA) • External induced voltage alteration (XIVA) • Seebeck effect imaging (SEI) • Dynamic • Soft defect localization (SDL) • Laser granulometry (Pye and Blott, 2004).
Semiconductor probing	• Mechanical probe station • Electron beam prober • Laser voltage prober • Time-resolved photon emission prober (TRPE) (Challinger et al., 2018)
Software-based fault location techniques	• CAD navigation • Automatic test pattern generation (ATPG) (Kezunovic et al., 2017)

5.4 Chemical markers in forensic chemistry

5.4.1 Forensic glass analysis

The main principle of forensic investigations or criminalistics is presence of actual trace evidence of a criminal or accidental activity which makes a record of itself. Vehicle hit-and-runs, housebreaking, robberies, stabbings, drive-by firings, and violence may produce glass pieces that could be utilized as evidence of association in criminal cases. Glass is produced by inorganic oxides. Main constituent of glass is silicon dioxide. Silica is designated as glass former while the carbonates and oxides are termed as additives (Landa et al., 2006). Overview of glass composition is explained in Table 5.2.

The aim of forensic glass analysis is to identify and distinguish the lineage of the glass and to determine if a glass fragment is linked to the crime scene or not. The glass analysis comprises physical and optical properties examination and testing for color, thickness of glass material, and glass density. If analysis of glass pieces from both amenities (i.e., fragments from crime location and from suspicious person's clothes) has similar density and refractive index, it does not mean that both are of one origin, but additional inquiry and examination is required such as elemental chemical analysis to confirm the exact composition of the glass material. Most often used methods are

Table 5.2 Overview of glass composition.

Formers of glass	Minerals	Rocks and compounds
Silica	Quartz and feldspar	SiO_2
Alumina	Nepheline and Aluminum hydroxide	Clay, Phenolite
Boron trioxide	Hydrogen borate and borax Sassolite borate mineral	Borate volcanic rocks
Phosphorous pentoxide	Phosphorus-based minerals	Phosphate rocks
Stabilizing agent		
Quick lime	Fluorospar, dolomite	$CaCO_3$ (lime stone)
Magnesia	Magnesite	Dolomite or magnesite source
Litharge, massicot	Sphalerite	Galena ores
Oxides of zinc	Zincite	Rock salt ZnO
Fluxing agent		
Sodium oxide	Sodium feldspar	Felsic igneous rocks
Potassium oxide	Phlogopite, diopside	Shale rocks
Lithium oxide	Spodumene	Lithia igneous rocks
Coloring agents		
Ferric, ferrous, sulfur, cobalt, nickel, chromium compounds to impart tinge or color		

microanalytical and spectrometric techniques. It includes probe microanalysis, neutron activation analysis, X-ray fluorescent technique, scanning electron microscopy, atomic absorption or emission spectrometry, and ICP-MS (Trejos and Almirall, 2010). Most of these techniques provide multicomponent elemental analysis. After determination of elemental composition, chemometric analysis is done followed by cluster and factor analysis procedures (Fig. 5.3 and Table 5.3).

Evidentiary value of glass can be explained by following points:

- Helpful in property offenses and crime investigation.
- Most easily available evidence on crime location because glass is abundant, stable, resistant against environmental factors and hence can be analyzed by the forensic with eloquent and reliable results even after enough time has passed between criminal offense occurrence and glass sample collection.

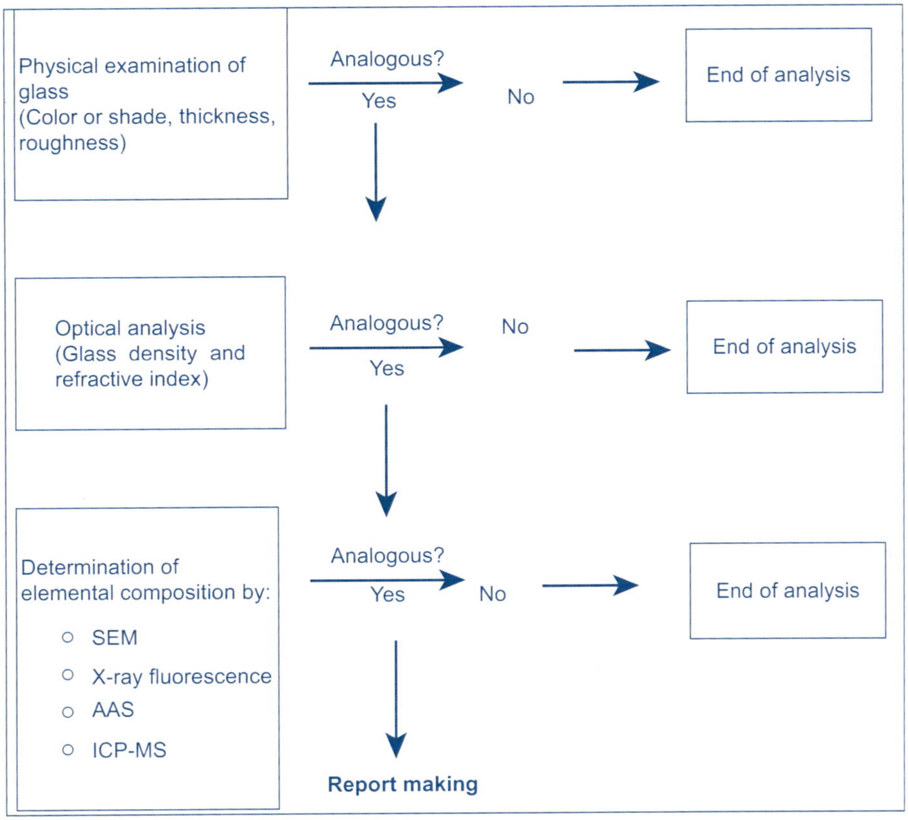

FIGURE 5.3

Outline for forensic glass analysis.

Table 5.3 Comparison of elemental detection techniques for glass analysis (Caddy, 2002; Carpi and Schweighardt, 2012).

Features	SEM/EDX	AAS	LA-ICP-MS	X-ray fluorescence
Nature	No	Yes	No	No
Limit of detection	Up to micron	1 g/L	0.01 g/L	Varies
Analysis range	6.1	3.5	8	5
Single/ multielement	Multielement	Single	Multielement	Multielement
Analysis cost	Moderate	Low	Very high	Moderate
Handling	Intermediate	Easy	Difficult	Intermediate
Sample requirement range	Milligram	milligram	Gram-milligram	Gram-milligram

5.4.2 Paint and coating analysis

Everything we encountered in our daily life has a coating or paint on its surface. So, paint can act as important evidence either the way it attaches to the surface or the mechanism it goes off with the suspect body in the form of chips or slurs. Forensic analysis divided paints into two main groups. First, coating with different paint layers following rigorous industrial processes applied usually on transportation means such as motor vehicle, aircrafts, sea boats, and luggage carriers. The other category demonstrates the paint and coating on household materials such as doors, windows, home tools, and walls. These paints are usually applied in only one-layer form. Ultimately, forensic laboratory can work on art objects for the purpose to identify anachronisms and forgeries. Paint evidence is usually obtained in the form of transfer when contact between coated surface and suspect occurs. Paint transfer quantity and quality depends on the characteristics of paint, nature of contact surface, and vividness of contact. Such transfer usually occurs in automobile accident when high-intensity friction force smears the paint from surface. Paint fragments are also formed by environmental conditions. The presence of control material is an important aspect in forensic paint analysis for comparison between suspect and paint surface.

5.4.3 Preparation of paint samples for evidence

A cross-sectional area of 5—15 micron is usually selected for paint sample preparation. It is usually cut by microtome sectioning technique followed by embedding procedures in an appropriate medium. An embedding medium should have following properties to meet the criteria (Derrick, 1995).

- Maintain at room temperature without shrinkage
- Nonreactive with sample
- Easily sectioned by microtome technique
- Examples: acrylic resins, waxes, epoxy resins, polyesters

The first and prime aspect of paint analysis involves optical analysis by stereomicroscopy and microscopy. The paint surface properties, such as number of layers, morphology, and color shade, are specific to each surface, either it is automobile paint or household item coating. Fourier transform infrared spectroscopy is the technique to determine organic and inorganic constituents, binders, extenders, and other additives. This technique is most widely used in forensic laboratory due to its reliability and fast procedure. In addition, Raman spectroscopy can be used. Pyrolysis, gas chromatography (GC), and mass spectrometry (MS) can be used in combination with

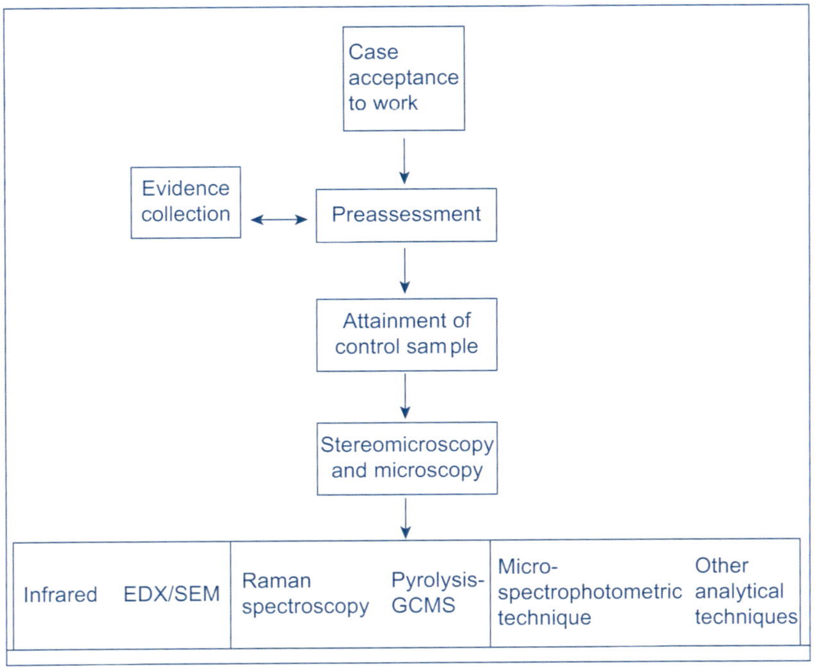

FIGURE 5.4

Outline for forensic paint analysis.

each other. Pyrolysis is the molecular breakdown of paint constituents by heat energy. Smaller fragments produced by pyrolysis are separated by GC and recognized or characterized by MS (Fig. 5.4 and Table 5.4).

5.4.4 Fibers and textile analysis

Fibers can generally be categorized as natural and man-made (Ko and Wan, 2014). Textile fibers are an important form of trace evidence as it can easily transferable (directly or indirectly) and have different forms. The transferred fibers are compared with the standard known fiber material upon testing. However, the time elapsed between incidence and evidence collection is an important factor as about 80% of material vanished after 4 h of transfer. So, it is mandatory for the investigator to gather all garments from the spot of incidence as early as possible. Of the other characteristics, color is an important aspect for comparison of fiber. Dyeing materials are used to color the fiber as dyes are able to absorb and reflect certain wavelengths of visible portion of light (Deedrick, 2000). The principal methods for the

Table 5.4 Techniques used for paint analysis with advantages and disadvantages (Chang et al., 2002).

Sr. No	Techniques	Advantages	Disadvantages
1	FTIR	Fast, reliable	Difficult interpretation
2	Raman spectroscopy	High spatial resolution No sample preparation	Fluorescence of sample
3	Pyrolytic GC-MS	High discriminating power	Destructive nature
4	Elemental analysis (SEM, EDX, X-ray)	Rapid and sensitive	Sometimes destructive
5	ICP-MS	Good precision and large linear dynamic range	Higher cost

identification and comparing of textile-dyed fibers depend on the principles of microscopy, spectroscopic, chromatographic, and mass spectrometric analysis. Microscopic analysis has become an important feature of textile fiber investigation (Grieve, 2000). A stereomicroscopic analysis provides various physical features of characterization like color, size, lustrous, and grooves or wrinkles in between fibers. FTIR analysis is widely used to characterize the type and nature of the fiber material (Fig. 5.5 and Table 5.5).

5.5 Case studies

Since material analysis in environmental forensic is a comparatively new field of study in Pakistan, very less data are available regarding its applicability during investigation. However, following is a brief description of cases in different regions of world to help the researchers in forensic analysis.

Case 1: A case involving robbery is reported in which a number of thieves entered a house and forced the occupants to open the safe. They tied one of the house members with a red cord and escaped. Police chased the thieves' vehicle from which several items were thrown out from the car window. During chase one of the thieves was arrested. Police found pieces of red cord on his gloves which found to match with the one they use to tie the home occupants (Schotman et al., 2017).

Case 2: A cyclist was found dead at the bank of a river. Police identified it as a hit-and-run case and collected 26 broken glass fragments from

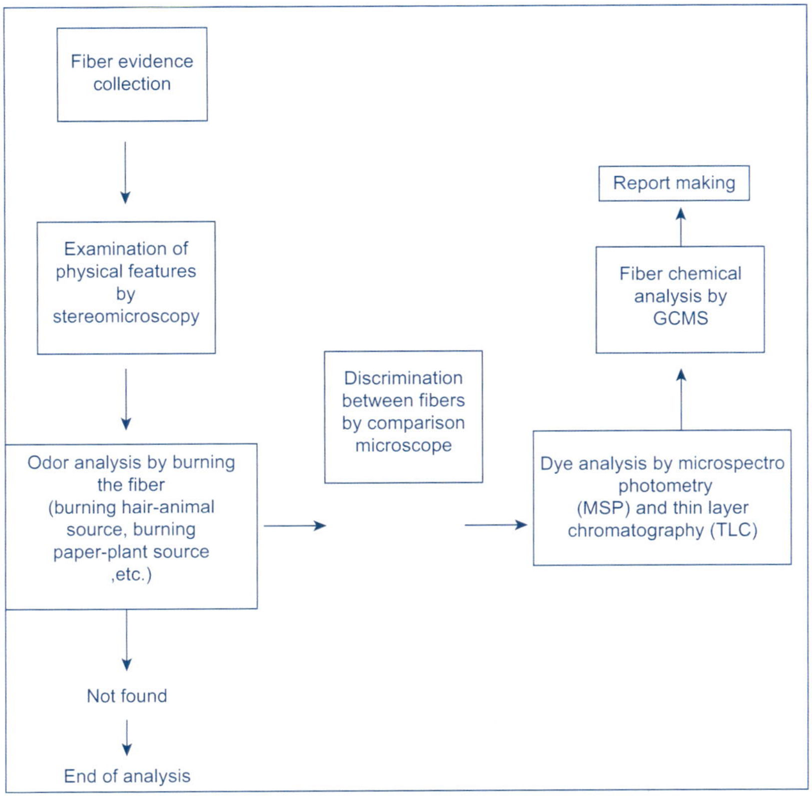

FIGURE 5.5

Outline for forensic fabric and textile analysis.

Table 5.5 Techniques for forensic textile analysis.

Techniques	Advantages	Detection limits
X-ray photoelectron spectroscopy (XPS)	Characterize surface of untreated textile fiber	Up to parts per million (ppm)
FTIR spectroscopy	Determine composition	1.6%–2.5%
LC-mass spectrometry	Identify and quantify a range of compounds	Up to 0.2 ppb
Separation techniques (TLC, HPLC)	Specifically, for separation of fiber dyes	Discriminate between closely similar dyes by elution time

accident scene. After few days, a suspected car stopped by police at check post whose headlights were broken. Three glass pieces were taken as control material from car. After investigation (i.e., tool marks examination and physicochemical analysis) of glass fragments, it was confirmed that the suspected vehicle took part in the hit-and-run case (Michalska et al., 2015).

References

Banas, K., Banas, A., Moser, H.O., Bahou, M., Li, W., Yang, P., Lim, S.K., 2010. Multivariate analysis techniques in the forensics investigation of the postblast residues by means of fourier transform-infrared spectroscopy. Analytical Chemistry 82, 3038–3044. https://doi.org/10.1021/ac100115r.

Brooks, C.R., Choudhury, A., 2002. Failure Analysis of Engineering Materials. McGraw Hill, New York, p. 233.

Caddy, B., 2002. Forensic Examination of Glass and Paint: Analysis and Interpretation. CRC press (Chapter 4).

Carpi, A., Schweighardt, A.J., 2012. Forensic Environmental Chemistry. In: Forensic Chemistry Handbook, pp. 1–22 (Chapter 1).

Challinger, S.E., Baikie, I.D., Flannigan, G., Halls, S., Laing, K., Daly, L., Daeid, N.N., 2018. Comparison of scanning Kelvin probe with SEM/EPMA techniques for fingermark recovery from metallic surfaces. Forensic Science International 291, 44–52. https://doi.org/10.1016/j.forsciint.2018.07.025.

Chalmers, J.M., Edwards, H.G., Hargreaves, M.D. (Eds.), 2012. Infrared and Raman Spectroscopy in Forensic Science. John Wiley & Sons (Chapter 5).

Chang, W.T., Chen, T.H., Yu, C.C., Kau, J.Y., 2002. Comparison of embedding methods used in examining cross-sections of automotive paints with micro-Fourier transform infrared spectroscopy. Forensic Science Journal 1, 55–60.

Clayton, J.L., 2017. Arc Jet Test and Analysis of Asbestos Free Solid Rocket Motor Nozzle Dome Ablative Materials. NASA Technical report Server, 20170005378.

Deedrick, D.W., 2000. Hairs, Fibers, Crime, and Evidence. US Department of Justice, Federal Bureau of Investigation (Chapter 2).

Demeestere, K., Dewulf, J., De Witte, B., Van Langenhove, H., 2007. Sample preparation for the analysis of volatile organic compounds in air and water matrices. Journal of Chromatography A 1153, 130–144. https://doi.org/10.1016/j.chroma.2007.01.012.

Derrick, M.R., 1995. Infrared micro spectroscopy in the analysis of cultural artifacts. Practical Spectroscopy 287–322 (Chapter 8).

Dunlop, S., Hayes, K., Leavy, P., Cusack, D., Maguire, R., 2017. An atmospheric pressure chemical ionisation liquid chromatographic–tandem mass spectrometry method for the analysis of benzodiazepines in urine. Journal of Chromatography B 1064, 22–27. https://doi.org/10.1016/j.jchromb.2017.08.023.

Ellingham, S.T., Thompson, T.J., Islam, M., 2018. Scanning electron microscopy–energy-dispersive X-ray (SEM/EDX): a rapid diagnostic tool to aid the identification of burnt bone and contested cremains. Journal of Forensic Sciences 63, 504–510. https://doi.org/10.1111/1556-4029.13541.

Fedick, P.W., Bills, B.J., Manicke, N.E., Cooks, R.G., 2017. Forensic sampling and analysis from a single substrate: surface-enhanced Raman spectroscopy followed by paper spray mass spectrometry. Analytical Chemistry 89, 10973–10979. https://doi.org/10.1021/acs.analchem.7b02798.

Grieve, M.I.C.H.A.E.L., 2000. Interpretation of fibres evidence. In: Forensic Examination of Fibres. Taylor & Francis, London, pp. 343–398.

Kezunovic, M., Esmailian, A., Govindarasu, M., Mehrizi-Sani, A., 2017. The use of system in the loop, hardware in the loop, and co-modeling of cyber-physical systems in developing and evaluating new smart grid solutions. In: Proceedings of the 50th Hawaii International Conference on System Sciences. https://doi.org/10.24251/hicss.2017.385.

Ko, F.K., Wan, Y., 2014. Introduction to Nanofiber Materials, first ed. Cambridge University Press (Chapter 5).

Landa, L., Landa, K.A., Thomsen, S.V., Hulme, R., 2006. U.S. Patent No. 7,037,869. U.S. Patent and Trademark Office, Washington, DC.

Michalska, A., Zadora, G., Świetek, M., 2015. Glass fragments as important criminalistic evidence–case studies. Problems of Forensic Sciences 102, 115–137.

Pye, K., Blott, S.J., 2004. Particle size analysis of sediments, soils and related particulate materials for forensic purposes using laser granulometry. Forensic Science International 144, 19–27. https://doi.org/10.1016/j.forsciint.2004.02.028.

Schotman, T.G., Xu, X., Rodewijk, N., van der Weerd, J., 2017. Application of dye analysis in forensic fibre and textile examination: case examples. Forensic Science International 278, 338–350. https://doi.org/10.1016/j.forsciint.2017.07.026.

Skoog, D.A., Holler, F.J., Crouch, S.R., 2017. Principles of Instrumental Analysis, third ed. Cengage learning (Chapter 4).

Trejos, T., Almirall, J.R., 2010. Laser ablation inductively coupled plasma mass spectrometry in forensic science. Encyclopedia of Analytical Chemistry 278, 338–350. https://doi.org/10.1002/9780470027318.a9103.

Weiszburg, T.G., Gherdán, K., Ratter, K., Zajzon, N., Bendo, Z., Radnoczi, G., Szakmány, G., 2017. Medieval gilding technology of historical metal threads revealed by electron optical and micro-Raman spectroscopic study of focused ion beam-milled cross sections. Analytical Chemistry 89, 10753–10760. https://doi.org/10.1021/acs.analchem.7b01917.

Further reading

Almirall, J.R., 2002. Elemental analysis of glass fragments. In: Forensic Examination of Glass and Paint. CRC Press, pp. 77–96.

Cho, L., Reffner, J.A., Gatewood, B.M., Wetzel, D.L., 2001. Single fiber analysis by internal reflection infrared micro-spectroscopy. Journal of Forensic Sciences 46, 1309–1314. https://doi.org/10.1520/jfs15151j.

Goodpaster, J.V., Liszewski, E.A., 2009. Forensic analysis of dyed textile fibers. Analytical and Bioanalytical Chemistry 394. https://doi.org/10.1007/s00216-009-2885-7.

Huang, M., Yinon, J., Sigman, M.E., 2004. Forensic identification of dyes extracted from textile fibers by liquid chromatography mass spectrometry (LC-MS). Journal of Forensic Sciences 49, 1–12. https://doi.org/10.1520/jfs2003295.

Meleiro, P.P., García-Ruiz, C., 2016. Spectroscopic techniques for the forensic analysis of textile fibers. Applied Spectroscopy Reviews 51, 278–301. https://doi.org/10.1080/05704928.2015.1132720.

CHAPTER

Forensic laboratory practices and quality

6

Irum Asif, MPhil, Marria Ghalib, MPhil, Shazia Iftikhar, PhD
Department of Environmental Sciences, Fatima Jinnah Women University, Rawalpindi, Pakistan

Chapter outline

Abstract

Over the past few decades, environmental managers and concerned legal practitioners had grown to count on forensics for the liability in environmental dispute resolution. Through scrutiny and scientific validation, good laboratory practices can be ensured. To reduce the possibility of invalidating the results, the chain of custody record is necessary for the integrity of the samples. In Pakistan, the Pakistan National Accreditation Council (PNAC) has been established under the administrative control of the Ministry of Science and Technology, Government of Pakistan, under the compliance of ISO/IEC 17025 standard. Forensic science is an emergent science that is truly recognized in Pakistan by the higher authorities. However, lot of work related to the databases and trained staff still lacking in

Trends of Environmental Forensics in Pakistan. https://doi.org/10.1016/B978-0-12-819436-2.00006-6

Pakistan cope with the current scenario of terrorism crimes scenes stored evidence. The success stories of using audio and video recordings by the Pakistan show a good initiative and meet the requirement.

Keywords: Chain; Laboratory accreditation; Laboratory practices; Pakistan; Quality assurance.

Graphical Abstract

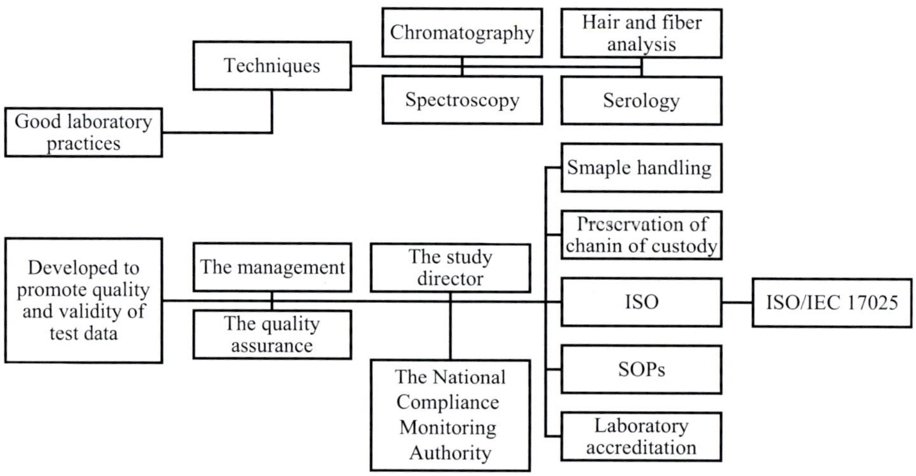

6.1 Introduction

Over the past few decades, environmental managers and concerned legal practitioners had grown to count on forensics as a means to allocate liability in environmental dispute resolution. Environmental forensics helps to determine whose contamination is it? When and how did the release occur?

After the 1970s' innovations and discoveries in analytical chemistry technology, investigators were able to analyze compounds selectively (both qualitatively and quantitatively). Simultaneously, quality assurance systems produce data that able to endure both scientific and litigation scrutiny and are admissible in court (Wait, 2001). Thorough scrutiny and scientific validation can be ensured through good laboratory practice (GLP).

6.2 Good laboratory practices

The principles of GLP have been established to endorse the quality and rationality of data used for defining the safety of chemicals and respective products. GLP is a quality system which aims to guarantee, through precise documentation, covering all aspects of a study and its environment, the quality, integrity, and reliability of safety data (Catalano, 2014a). GLP is principally a managerial concept covering the process and the conditions under which laboratory studies are planned, performed, monitored, recorded, and reported.

GLP is centered on the following four pillars to achieve the desired goals:

1. The management
2. The quality assurance
3. The Director (Study Director)
4. The National Compliance Monitoring Authority

Among these, quality assurance helps for in-house control of principles of GLP, whereas National Compliance Monitoring Authority accommodates international recognition and acceptance of acquired data. Structure of GLP is sustained by all four pillars. All these pillars are vital for sound functioning and accurate monitoring to achieve good-quality data. Undoubtedly, other aspects are almost equally important, but quality assurance, National Compliance Monitoring Authorities, and Study Director hold the key positions.

6.2.1 Sample handling and integrity maintenance

Science plays an important role as part of an effective investigative response to any criminal happening by providing skills to analyze biological and associated signs in the evidence. The collection and preservation of forensic evidence are vital to effective investigation. If evidence is not properly collected; degrades; or contaminated during collection, handling, transport, or storage, the characterization and analysis may be bargained.

It is recommended for collecting biological evidence to not remove the stain from an object. It is preferred to collect the object with the stain. Depending upon the nature of the evidence, a stain can be secured by immobilizing the evidence item in a cardboard box or by taping a piece of paper over the stain. Another option is to remove the stain by cutting it out, e.g., from a piece of cloth or carpet. The recommended method for swabbing a stain is to use a minimum amount of distilled water to dampen the clean substrate (e.g., cotton swab or cotton gauze) and then absorb the stain onto the slightly dampened substrate. Cotton swabs have traditionally been used to collect biological samples. Alternately, cotton gauze, which does not have any additive (e.g., fabric sizing), is also used. Drying biological samples is crucial in preservation of samples, as moisture is necessary in most of the biochemical reactions that result in degradation. Saliva samples placed onto swabs placed in containers with holes show more degradation and results in less DNA yield as compared with saliva samples placed on swabs packed in paper envelopes. It is preferred to air dry the samples before packing the samples in paper envelopes. In humid environment, use of desiccants is also preferred.

One last consideration is to minimize the contamination of biological sample with potentially deleterious material. It is done by collecting the sample with minimum amount of surrounding material. For example, cigarette ash contamination results in failed DNA test. Therefore, it is advised to collect cigarette butt(s) from ashtrays without the ashes. This also applies to collecting biological samples from soil with a greater chance of a successful DNA test program with least soil matrix (Budowle et al., 2006). Once the sample is collected, the most important step is to label the sample with the basic information as shown in Fig. 6.1. Some precautions are kept in consideration while labeling the sample, e.g., use of indelible ink for labeling that is unaffected by the gases and temperatures to which it will be subjected. Alternatively, other methods of identification can also be used, e.g., bar coding.

6.2.2 Sample collection

To maximize the validity of the results, the collected samples must be carefully removed from the monitoring device, placed in nonreactive containers,

(Name of Sampling Organization)
Sample ID No: ____________________
Storage Conditions: ____________________
Sample Type: ____________________
Date/Time Collected: ____________________
Site Name: ____________________
Site Address: ____________________
Sampler Signature: ____________________

FIGURE 6.1 Representation of sample label.

properly labeled, and sealed. The sample label must be properly attached to the container so that it cannot be accidently removed. It is suggested to use tamper-evident custody seals in certain high-profile cases. Custody seals on sample containers provide visual evidence regarding the tempering or opening of the container. It is also suggested for the operating technicians to duly sign the label of the sample. In some cases, unauthorized personnel, such as transportation security administration officers can inspect the samples, and the proper use of custody seals minimizes the loss of samples and provides direct evidence whether sample containers have been opened and possibly compromised.

6.2.3 Packaging materials

After the sample has been collected, it should be protected against loss and contamination by the use of proper packaging material. This packaging can also be used to record the chain of custody (CoC) and to describe the evidence item. Mostly paper packaging is used, being the porous material allowing air exchange with the atmosphere. It helps in complete drying, even when the sample is not entirely dry at the time of initial packaging. In situations when the sample is a heavily bloodstained garment, it should be allowed to air dry before packaging or in packaging. Best method is to roll the cloth in a clean piece of paper to avoid the direct contact between separate stains. Ideally, each evidence item should be packaged in a separate paper container. Appropriate size of packaging bag is recommended to avoid unnecessary folding of the evidence. The packed samples are subjected to long-term storage conditions for validation in some situations.

6.2.4 Storage and archival

Samples must be properly held to ensure that it is free of contamination and the results generated are actually of the samples collected under the same conditions. For this reason, they should be kept in a secured location. In situations when the sample is not under the direct control of the sample custodian, it can be preferably placed with limited access during transportation and storage in locked vehicle, refrigerator, or laboratory. The samples must be properly stored and not discarded until the case of investigation and legal action is in process after the analysis. The security measures should be written and signed by the handlers of the sample on the CoC form in a roper format. It is suggested that the samples should be properly archived the first year in cold conditions (e.g., at 4°C) and for the next 2 years at ambient conditions.

6.2.5 Chain of custody

A written record of the detailed locations of sample and dates must be accompanied by the sampling program as evidence. Similarly, the CoC record is also necessary showing of the integrity of the samples. Procedures may vary, but an actual CoC record sheet with the names and signatures of the receivers works well for tracking physical samples. The samples should be handled by minimum concerned persons only associated with the monitoring program. A general rule "the fewer hands the better" works well till the sample is properly sealed without affecting its integrity. CoC must be properly narrated with the details of the state of evidence, collected from whom and when. A CoC form must be attached to it and can be used to track the handling of the samples through different stages of storage, processing, and analysis. An example of a form that may be used is shown in Fig. 6.2. When using professional services to transport physical samples, only reliable registered services that provide a tracking number should be used. Information labeling the enclosed samples should be placed on the bill of lading.

A copy of the shipping receipt and tracking number should be kept as a record. The person receiving the evidence must be authorized and properly addressed on the package. A procedure must be in place to ensure that samples are delivered to the appropriate person without being opened or damaged. In this circumstance, the sample is considered still in transport until received by the authorized addressee. It may be necessary to ship and/or receive samples outside of normal business hours. A procedure should be developed in advance that considers staff availability, secure storage locations, and appropriate storage conditions (e.g., temperature-controlled).

6.3 Techniques

Given the success of forensic science in providing realistic, testimonial evidence for the courts, this discussion highlights the view of traditional methods used in forensic science. The purpose is to help us understand how we can apply similar techniques when dealing with information systems. Classical forensic analysis methods include the following:

- Spectroscopy, chromatography, hair and fiber analysis, and serology (such as DNA examination).
- Odontology, pathology, structural engineering, anthropology, toxicology, and examination of questionable documents.

Chain of Custody		
Organization:		
Address:		
Contact Details:		
Field Sampler:	Signature:	
Date of Collection:		
Relinquished by:	Date /Time	Received by:
Remarks:		
Relinquished by:	Date /Time	Received by:
Remarks:		
Relinquished by:	Date /Time	Received by:
Remarks:		
Relinquished by:	Date /Time	Received by:
Remarks:		
Relinquished by:	Date /Time	Received by:
Remarks:		
Relinquished by:	Date /Time	Received by:
Remarks:		
Relinquished by:	Date /Time	Received by:
Remarks:		
Relinquished by:	Date /Time	Received by:
Remarks:		
Relinquished by:	Date /Time	Received by:
Remarks:		
Relinquished by:	Date /Time	Received by:
Remarks:		
Relinquished by:	Date /Time	Received by:
Remarks:		
Relinquished by:	Date /Time	Received by:
Remarks:		

FIGURE 6.2 A sample representation of chain of custody form.

- Behavioral patterns revealed by tests, such as polygraphs and psychological exams.

Most of these forensic disciplines began to flourish alongside the evolving science of criminalistics, which, in the United States, emerged during the 1920s. Advances in medicine, microscopy, and chemistry pave the path for the implementation of scientific analysis rather than pure observation and eye witness as the cornerstone of criminal investigation. The result of these advances was to replace supposition with reality and present testimonial evidence to the judge or jury in criminal or civil proceedings (Eckert, 1996).

It is evident that the vast majority of analytical methods employed by traditional forensic science grew out of university laboratories. In fact, before 1929 no official crime laboratory existed in the United States. Instead, police departments solicit the assistance of prominent university professors to help them collect and examine potential evidence in using scientific analysis in crime solving. Professionals with particular interest in the forensic aspects of analysis began to shift their practices to newly established laboratories that focused on forensic analysis in support of the courts.

The gradual prototype shift, from instinct to fact derived from analysis, took hold in the early 20th century for a number of reasons. The sciences, both hard/physics and soft/biology, are advancing speedily, and many of the discoveries are being out to a larger percentage of the common population. Perhaps more important was the fact that alone observation had been proven to eventually leading to suspect conclusions. Over time, scientific evidence in the courts became subject to more rigorous scrutiny as crime evidence. Individuals in the court system realized that testimony submitted as scientific and conclusive was beyond their complete understanding. With passage of time, it became clear that these analytical methods are undisputable in courts. They were derived by experimentation that contained measures of error and other guides to help describe the accuracy of statistics and narrative results. This concept led to the development of standards that must accompany scientifically derived testimonial evidence.

Mostly in criminal proceedings, the scientific methods are considered as the most relying evidences in courts and public opinion. Perhaps the most commonly stated is DNA profiling, yet it is least understood. This relatively new method is performed by forensic serologists for the courts. It gains importance due to the purported ability to discriminate down to the level of the individual, thus replacing older methods like blood typing as a primary evidentiary mechanism.

Looking a little deeper, DNA analysis, though certainly more reliable than blood typing alone, is not a solution. The general perception is that the DNA evidence in court is unquestionable and cannot be challenged. It is based on the probability that the DNA analysis will correctly determine that a defendant was the source of evidence found at the crime scene. When collecting statistics considering laboratory practice and data collection factors, false-positive detection rates range from one per hundred to one per thousand. It is therefore believed that the studies based on population genetics can potentially become unrelated since so much error can be exclaimed by incorrect collection and handling of the DNA source material. Although the scientific community has agreed that DNA profiling is very accurate, the reliability of any particular test still remains an issue. This issue is focal for current and future practitioners in digital forensic analysis.

Until recently, the scientific community has been noticeably absent from the development of standard procedures and processes related to digital forensic analysis. This trend has led to court dependence on pattern rather than statistical significance when ruling on acceptability of evidence derived from digital sources (Carrier, 2002). As juries, judges, defense attorneys, and asset managers become well informed in digital technology and understand its complexity, it is likely that a more rigorous approach to digital forensic analysis will be needed.

Once this practice begins, the jury will ask more compelling questions and scientifically proven explanations from those providing testimony. This new view of evidence will force a paradigm shift that will slowly change law enforcement's use of technology in business, industry, government, and the military. For the digital forensic analyst working in near-real-time environments, it will result in quicker responses based upon more reliable evidence that is derived from established technology. It is believed that digital forensic analysis is at the same stage as the other disciplines in the early part of the 20th century. It is an evolving scientific discipline that is becoming more familiar to the general public.

Measures of reliability and accuracy for the analytical techniques reach the level of confidence expected in the testimony. Information resulting from computer forensic analysis has yet to be questioned extensively by defense lawyers, and computer misuse by the analysts in still not observed. Most techniques used today are assumed plausible because they are developed by trustworthy companies, used by practitioners in the field, and have been used previously in courts or other settings to persuade authorities. Also of note is that the techniques and the conclusions they produce have yet to be tested for reliability in controlled environments under experimental protocols. Strict interpretations of the rules of evidence and court precedence suggest that this digital forensic evidence testing will be a vital part in future. Some of the methods employed by the traditional forensic sciences have much to teach the new comers in the field of digital forensic analysis. The emphasis is in the scientific processes at work as the discipline evolves. Using DNA as an example, a number of discoveries were made in the span of a number of years. In 1869, Johann Meischer's analysis of old bandages used in the Crimean War led to the discovery of nuclein. In 1953, Watson and Crick defined the structure of DNA, which gave researchers a blueprint. In 1977, Gilbert and Sanger defined DNA sequencing allowing researchers to analyze small parts of the structure. Alex Jeffery found the human part of the DNA strand in 1985 making unique comparisons possible. It is expected to see the combined evolutionary techniques of serology and computer forensics with strong interaction among academic researchers, field practitioners, and legal experts.

Despite the limited forensic capability, there is minimum obligation on prosecution to prove their recovery items by forensics. This implies showing that a certain object, for instance, is a live hand grenade or the "powder" seized is actually explosive in nature. These basic views are also not complied in many cases, but where chain of evidence is consistent, reports of convictions are present. However, investigators do not seem to comprehend that they have to "forensicate" a bombing crime scene, and thus gross irregularities occur. These could be as basic as explosive materials not recovered as evidence in bomb explosion cases. Other types of modern evidence are also given little importance. Due to sociocultural constraints and religious views, DNA evidence collection and postmortem especially of females is still considered a taboo in Pakistan and seems to have a hard run in Pakistan. This was highlighted in a rape trial, when a criminal court rejected a DNA report incriminating suspects, while acquitting them based on eyewitnesses evidence. This is the traditional mindset of the courts on the sovereignty of eyewitness evidence, even to the rejection of DNA evidence. It is believed that the trends seem to be changing notably due to the use and implementation of state-of-art forensic laboratory in Punjab.

6.4 Laboratory accreditation

GLP sustains quality assurance and vice versa. Quality assurance is ensured through accreditation of laboratories involved in forensic investigation.

Laboratory accreditation is a means of determining the technical competence of research lab to perform specific case of testing, measurement, and calibration (Yingwa et al., 1999). Laboratory's accreditation benefits both laboratory and users of laboratory services and the general public. Testing and calibration laboratories add a great deal from a technically sound judgment and accreditation by an internationally recognized accreditation body including the following:

- By accreditation process, a laboratory gains by necessarily building a quality management system which functions to reduce procedural errors.

- It also helps to provide a steady checkup that helps a laboratory's management to create continual improvements in its operation.

In Pakistan, Pakistan National Accreditation Council (PNAC) has been established under the administrative control of the Ministry of Science and Technology, as an agency to accredit conformity assessment bodies such as laboratories and certification bodies. PNAC accredited first laboratory in 2004. Being a pioneer member of the Mutual Recognition Arrangement (MRA) of the Asia Pacific Laboratory Accreditation Cooperation (APLAC) and the International Laboratory Accreditation Cooperation (ILAC), testing reports issued by PNAC-accredited laboratories are acknowledged globally by MRA members spanning more than 70 countries (Table 6.1).

6.5 ISO

6.5.1 ISO/IEC 17025—the standard for laboratory competence

Universal requirements for laboratory competence are described in the ISO/IEC 17025 standard (Catalano, 2014b). ISO/IEC 17025 institutes a global baseline for the laboratory accreditation of all sorts (Fox, 2003). Recognition of such competence generally requires that laboratories obtain accreditation comprising on-Internet site and operation assessments as well as ongoing technique testing. Assessment of competence requires persons who not only understand the necessity of the banner but have a sufficient depth of understanding about the specified mental testing to make judgments of competence. The assessors must apprehend the principles underlying the prerequisite of ISO/IEC 17025, which are not always obvious. Subterfuge adherence by a laboratory to the requirements of the standard, while better than no system at all, is not an approach path that instills self-assurance in its ability to produce valid test resultant role.

6.6 SOPs

Standard operating procedures (SOPs) are envisioned to describe procedures customarily employed in the performance of test facility operations. Without a doubt SOPs are "documented procedures which describe how to perform tests or activities normally not specified in detail in study plans or test

Table 6.1 List of accredited laboratories in Pakistan.

Laboratory	Accreditation granted	ISO requirements	Materials/products tested	Type of test/ property measured
Firearms and Toolmarks Testing Laboratory Punjab Forensic Science Agency (PFSA), Lahore	02-11-2017	ISO/IEC 17025:2005	Cartridge cases Firearms Victim clothing/ vehicles	Comparison test Functionality test Gunshot residue/ distance determination test
Chemical and Environmental Laboratory, SGS Pakistan Private Limited, Karachi, Pakistan	06-06-2006		Chromium ore	Chromium
Centre For Environmental Protection Studies (CEPS), Pakistan Council of Scientific and Industrial Research (PCSIR) Laboratories Complex, Lahore, Pakistan	24-08-2006		Water and wastewater Rice	Determination of oil and grease Determination of phenolic compounds Determination of cadmium in rice Determination of lead in rice
Laboratory of Green Crescent Environmental Consultant (Pvt) Ltd., Lahore, Pakistan	02-08-2017		Water and wastewater	Chloride Oil and grease

guidelines." SOPs define all steps involved in the performance of an activity in a comprehensive way.

6.7 Institutes in Pakistan and gaps in their performance

DNA evidence has been gaining traction in Pakistan. However, no comprehensive database of DNA currently exists in Pakistan to match samples from terrorism crime scenes against the stored evidence. Also, the DNA analysis of the collected evidences is time-consuming, and the courts are sometimes hurried. And the reports required from the remote areas which do not have developed forensic capability reach courts even later. A ray of hope comes in the form of the Punjab forensic science authority for the analysis of most of the DNA evidences. Set up in the location of Lahore, this laboratory has become the gold standard of forensic examination in Pakistan. Diligently maintained facilities with state-of-the art equipment, foreign-trained scientists, and proper accreditation and control procedures in place, the PFSA is indeed admirable. However, with the passage of time, there seems to be excessive number of samples, and the laboratories seem to be overburdened. As the USIP report highlights, undercurrents of capacity gaps of the police in handling modern evidence currently run deeper than even one of the best laboratories is the world, the PFSA, can cure. Police are innately cynical about the usage of modern devices and methods of investigations but still rely on premised in no small part by the widespread usage of ocular evidence. In the modern age, our courts still rely on eyewitness evidence; the signal that the law enforcement presumably gets is that such evidence sways supreme, to the exclusion of any other type of evidence.

The dilemma is that policing has never been reorganized on modern lines, and it remains largely oblivious to the need for such. CoC of forensic evidence is not satisfactory, with evidence either not being sealed properly or kept with police pending dispatch. Then there are some cases when the evidence is actually lost in police custody. Pakistani citizens have stronger desires for DNA profiling, and the government of Pakistan is trying to develop a national DNA database of all its citizens. The advantage of this DNA database will be to link it with the currently available biometric-based computerized National Identity Card (NIC) system to search out criminals and terrorists and finally to identify suicide bombers and victims of accidents, i.e., man-made as well as natural disasters and military conflicts (Table 6.2).

Table 6.2 List of high profile cases solved by the audio video department of Punjab Forensic Science Agency.

Sr. No.	Case	Year
1	Jaranwala Case	2018
2	Kasur Zainab Case	2017
3	Quetta Case	2016
4	Model Town Case	2014
5	Kasur Case	2017
8	Sangla Hill Lip Stitching Case	2016
9	Sehwan Case	2017

6.8 Case studies

Recently Punjab government in Pakistan has established the world's second largest forensic laboratory (Punjab Forensic Science Agency) in Lahore to counter terrorism and help police department in the investigations of criminals. This forensic laboratory has 14 different forensic disciplines, including DNA and serology, toxicology, pathology, forensic photography, computer forensics, narcotics, death scene investigation, crime scene investigation, trace chemistry, firearms and tool marks, latent prints, questioned documents, polygraph, and audiovisual (Zar et al., 2013). The government of Punjab is also trying to make collaboration with different renowned forensic institutes of the world to get their support for strengthening and development of the laboratory. There are still so many gaps and flaws in DNA profiling in Pakistan. These are lack of experts in this field; inadequate funds and CoC; lack of awareness, training, and equipment; lack of communication between forensic DNA analysts with forensic DNA researchers. Pakistani DNA analysts need special training and are not updated with the current DNA research work in forensics. They need to keep communication and collaboration with different forensic research institutes within country and out of country. Efficient and well-functioning forensic DNA laboratories are necessary in Pakistan, particularly in Karachi, Balochistan, FATA, and Khyber Pakhtunkhwa province of Pakistan to overcome crimes and terrorism. There are only few institutes in Pakistan which have started research work in DNA forensics, such as Fatima Jinnah Women University, Rawalpindi; Center of Excellence in Molecular Biology (CEMB) University of the Punjab, Lahore; University of Veterinary and Animal sciences, Lahore; and Government College University, Lahore, Pakistan.

References

Budowle, B., Schutzer, S.E., Burans, J.P., Beecher, D.J., Cebula, T.A., Chakraborty, R., Cobb, W.T., Fletcher, J., Hale, M.L., Harris, R.B., Heitkamp, M.A., Keller, F.K., Kuske, C., LeClerc, J.E., Marrone, B.L., McKenna, T.S., Morse, S.A., Rodriguez, L.L., Valentine, N.B., Yadev, J., 2006. Quality sample collection, handling, and preservation for an effective microbial forensics program. Applied and Environmental Microbiology 72 (10), 6431–6438.

Carrier, B., 2002. Open Source Digital Forensics Tools: The Legal Argument. stake, pp. 1–11.

Catalano, T., 2014a. Forensic chemistry. In: Good Laboratory Practices for Forensic Chemistry. Springer, Cham, pp. 7–10.

Catalano, T., 2014b. International Organization for Standards (ISO). In: Good Laboratory Practices for Forensic Chemistry. Springer, Cham, pp. 41–44.

Eckert, W.G., 1996. Introduction to Forensic Sciences. CRC Press.

Fox, A., 2003. GLP regulations vs. ISO 17025 requirements: how do they differ? Accreditation and Quality Assurance: Journal for Quality, Comparability and Reliability in Chemical Measurement 8 (6), 303.

Wait, A.D., 2001. Environmental forensic chemistry and sound science in the Courtroom. Fordham Environmental Law Journal 12 (2), 293–327.

Yingwa, S.H.E.N., Lisai, S.H.E.N., Hong, Z.H.O.U., Ling, L.U., Jinglei, N.I.E., Fu, S.O.N.G., &Luxin, W.A.N.G., 1999. A study on the assessment criteria for chemicals testing laboratories in compliance with GLP. Research of Environmental Sciences 1, 1–10.

Zar, M.S., Shahid, A.A., Shahzad, M.S., 2013. An overview of crimes, terrorism and DNA forensics in Pakistan. Journal of Forensic Research 4 (4), 1–2.

Further reading

Koehler, J.J., Chia, A., Lindsey, S., 1995. The random match probability in DNA evidence: irrelevant and prejudicial? Jurimetrics 201–219.

CHAPTER

Bioterrorism, an Emerging Threat

7

Zainab Naeem, MPhil, Nageen Sohail, MPhil, Shazia Iftikhar, PhD

Department of Environmental Sciences, Fatima Jinnah Women University, Rawalpindi, Pakistan

Chapter outline

Abstract

Various books such as *Handbook of Viral Bioterrorism and Biodefense* (Clercq and Kern, 2003), *Biosecurity and Bioterrorism: Containing and Preventing Biological Threats* (Ryan and Glarum, 2008), and *Bioterrorism: Threats and Deterrents* (Çankaya and Kibaroğlu, 2010) have discussed threats, causes, and effects of bioterrosim in different parts of the world. The chapter on "Bioterrorism, an Emerging Threat in Pakistan" focuses on the current situation of bioterrorism in Pakistan. Several examples from all around the world have been quoted to highlight different incidents of bioterrorism across the world. After addressing vicious epidemics, several laboratory methods for detection and surveillance for bioterrorism have been elaborated. The case studies of bioterrorism in terms of warfare highlight use of bioweapons at political fronts. Similarly, the success story of controlling bird flu and advent and persistence of dengue show the status of bioterrorism and Pakistan's preparedness to deal with this menace. Pakistan has strengthened itself to deal with bioterrorism by adopting different national and global tools for governance. However, the rapid decline in environmental quality leads to an increase in biohazards. It

Trends of Environmental Forensics in Pakistan. https://doi.org/10.1016/B978-0-12-819436-2.00007-8

is important to address bioterrorism at every forum so that sustainable solutions can be devised to tackle environmental deteriorations.

Keywords: Biohazards; Bioterrorism; Environment; Pakistan; Warfare.

Graphical Abstract

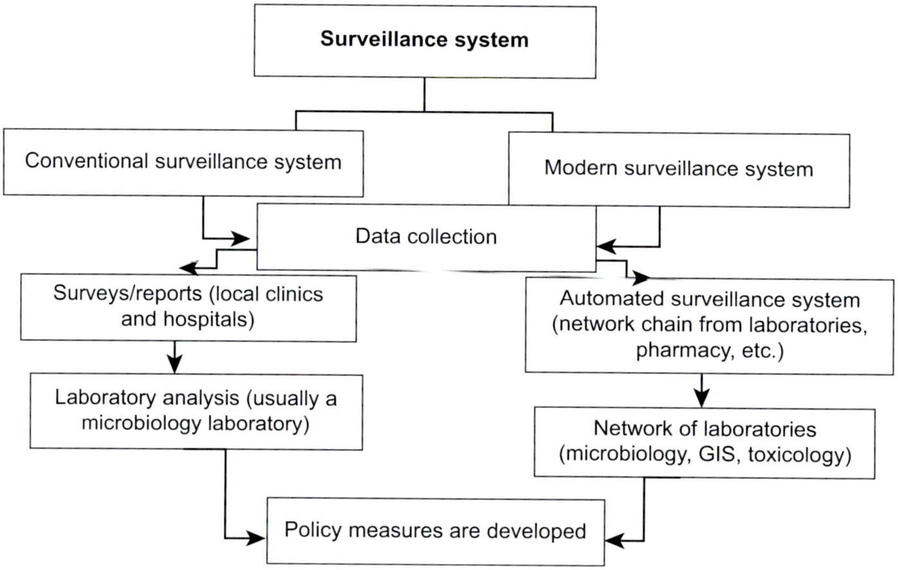

7.1 Introduction

Bioterrorism, bioweapons, or biowarfare is a term used to describe the intentional or deliberate use, spread/release, dissemination or production of living organisms, toxins, and chemicals of animal or plant origin to produce such diseases that can harm or cause death of humans, animals, plants, and even the overall environment. The terrorist can spread it via air, water, soil, and food, and they can take few hours to several days to cause the effect. It is a very broad subject that includes diverse range of organisms (bacteria, pathogens, and viruses) and chemical toxins which are used to spread terror, danger, risk, and threat. This can be performed by an individual having vested interests or for retaliation, but it can also be sponsored by some government as part of a political agenda. In earlier days, anthrax and small pox diseases were used as a source of bioterrorism in the West and even recently, anthrax is a very commonly used agent for bioterrorism. This form of terrorism is the easiest because the agents are available easily, do not cause sound or blast (unless it involves chemicals), can cause harm across borders, and are widespread mostly as it can wipe out entire population (Khadori, 2006). There are three broad categories which specify the term bioterrorism. The categories are based on the threat ranging from highest to moderate and eventually to gradual (A, B, and C) (Table 7.1).

Category A consists of those disease-causing agents which can be easily released and can spread from one person to another. They can become a public health problem and result in socioeconomic impact because people can become ill and do not go to jobs, schools or handle business which ultimately results in social impact. Some examples are bacteria such as *Bacillus anthracis* (anthrax-causing bacteria) and *Yersinia pestis* (plague-causing specie), viruses such as filoviruses (cause Ebola), and toxins such as *Clostridium botulinus* (botulism-causing toxin).

Category B includes those agents that are moderately easily transmitted and do not cause deaths, but some people are affected. For instance, bacteria specie such as *Salmonella* (food contaminant) and virus such as alphaviruses (cause encephalitis).

Category C includes those pathogens that are being genetically engineered to cause effect in future and they have a potential to result in high death toll. For example, *Naegleria* and Nipah viruses.

Historically, bioterrorism has been termed as the earliest form of terrorism to be practiced since pre-Christian era when wars were common and technology was not evolved as it is today. However, world was making advancements in science and methods were being invented to take over the world and crush the enemy in a shorter time. So the easiest way to do it was to spread diseases that can weaken the enemy. It is usually believed that the plague epidemic in Europe was caused by some sailor who carried the disease from the enemy country (Riedel, 2004). Similarly, there has been

Table 7.1 Distinction between categories of bioterrorism.

	Category A	Category B	Category C
Level	Highest	Moderate	Gradual (future use)
Impact	Large population	Small population	Potential to effect large population
Examples	Bacteria (*Bacillus anthracis*) Virus (filoviruses)	Bacteria (*Salmonella*) virus (alphaviruses)	Virus (Nipah virus)

incidences when during election, a poisonous dose of bacterial agent was mixed in the salads of local restaurants so that the voters turn out will be less and the competitors will lose (Barras and Greub, 2014). Then later, it became common to use bioterrorism during world wars, when it assumed the shape of chemical warfare. Mustard gas was used to blind the enemy, poison used to be sprayed in the freshwater sources to kill the population, food used to be poisoned, viruses were spread through prostitution, and so on. Recently, Ebola virus, bird flu, rota virus, H1N1, influenza, Naegleria, dengue fever, and many other forms of diseases are being spread among those countries where population is not getting controlled such as African and some Asian countries. Also, in Syria, recently the President Bashar al-Assad bombed the civilians with chemicals which resulted in huge casualties of children. It has been reported that after the onset of Gulf War, many countries sent agents in African countries to acquire such viruses and bacterial agents such as Ebola and botulinum to cause harm to either the enemy or the groups that were engaged in uprising against the tyrannical governments. Even such chemical gases like nerve gases are available in all European countries to counter the war threat (Handerson, 1998). Some examples of bioterrorism from history are enlisted in Table 7.2.

7.2 Laboratory detection and surveillance methods for bioterrorism

Globally, there are laboratories established to address bioterrorism. According to the Centers for Disease Control and Prevention (CDC), there should be a laboratory response network which should include clinical microbiology laboratories that are responsible for identifying, detecting, and reporting bioterrorism. In addition, they must suggest the measures that can be adopted to remedy the effects. Moreover, the laboratory response network can be grouped category wise, dealing with the issue, like a category A laboratory can be a primary laboratory that can help detect such agents which can likely cause bioterrorism effects and so they should be equipped to conduct primary analysis. The B category laboratories can help to isolate and identify the exact agent responsible for bioterrorism and work for its remedial measures and design such strategies. Similarly, the category C laboratories should work to identify those agents that can cause potential damage as a bioterrorism agent in future; they all must act as public health laboratories. Mostly, microbiologists are capable enough to work out the effects and remedies for bioterrorism and the population suffering or at risk to suffer from its

Table 7.2 List of biodiseases around the world.

Diseases	Year	Origin/ country	Reported causalities	Source
Black death pandemic	13–15th century	Europe	25 million reported deaths	Shinwari et al. (2014)
Smallpox virus	1600s	North America	400,000 causalities	Riedel (2004)
Severe acute respiratory syndrome (SARS)	2002–03	Hong Kong/ China	298	Mikes (2009)
Avian influenzas	2006	Azerbaijan, Cambodia, China, Egypt, Djibouti, Indonesia, Iraq, Laos, Thailand, Turkey, Nigeria, and Vietnam	150 million affected birds resulted in 335 confirmed cases	Brooks (2007)
Equine influenza	2007			
Swine flu	2009/ 2014	Worldwide	203,000 deaths	Gholipour (2013)
Tuberculosis	2016	Worldwide	10.4 million affected people	CDC (2018)

impacts. However, toxicologists also have a significant role to offer in this regard, as they are the ones who can identify the toxins from food, air, water, and even soil sources that can have an impact on the entire population and pose a risk to the overall public health (Lim et al., 2005). Usually, blood samples are taken from humans and animals for the identification of the cause, while water, crop, soil, and air sampling is also performed to identify the source. Ideal sampling techniques have been recommended by CDC which should be adopted globally as a standard procedure. Some common globally used laboratory techniques to detect bioterrorism agents include centrifugation, filtration for the separation of microorganism on cellular level, dielectrophoresis, immune-magnetic separation; and mostly polymerase chain reaction is also utilized by laboratory experts to identify biowarfare agents that can have impacts on genes and at molecular level (Bravata et al., 2004).

Before the laboratory work begins, there is a surveillance system that includes collection of data either through observation or surveys and later reporting it to the laboratories. The data should be collected from local clinics and hospital reports because bioterrorism is a public health issue, so a nation-wide surveillance system should be devised to monitor and manage the issue. This kind of surveillance system is also termed as conventional or traditional surveillance; however, modern surveillance system has also been adopted by many countries in the world to counter bioterrorism which include nation-wide automated surveillance for disease-related syndromes, analysis of routinely collected clinical, administrative, pharmacy, and laboratory data. Another important thing that should be considered for surveillance is the cost, social barriers, and environmental ethics. So, it does not matter which surveillance type is adopted as it depends upon the country's environmental, economic, and social conditions. Furthermore, there has to be a system of proper management and chain of command which should be supervised and regulated under the government as it is a concern for entire population. A regulatory body should be established to further remove the bias that can be questioned later (Hutwanger et al., 2003). However, problems can still exist because it is not easy to identify the culprit who release or transmit agents because the recipes to spread such kind of terror are easily available and accessible through the Internet. Also, the environment ethics is the subject being ignored the most specifically by the scientists who are engaged in performing experiments on viruses and bacteria, genetic modifications, etc. and is becoming a reason for the wide spread of bioterrorism in the world (Fig. 7.1).

7.3 Bioterrorism and its impacts on environment of Pakistan

Biosafety is a well-established concept which gained fame after the incident of 9/11. Similarly, despite Pakistan striving to survive in the global dynamics, climate change, however, is one of the biggest forms of bioterrorism in Pakistan. Massive desertification in agricultural plains, diverge droughts in Thar, massive flooding in arid and semiarid regions and irregular pattern of monsoon are the results of climate change (Nazar, 2016). Gayari Sector avalanche resulted in the death of 121 Pakistani soldiers. Pakistan witnessed its worst history of floods in 2010 which caused damage to over 20 million people, killed more than 2000 people alone, and resulted in huge economic loss to the country (Jaspal, 2010). Flash floods caused by torrential monsoon

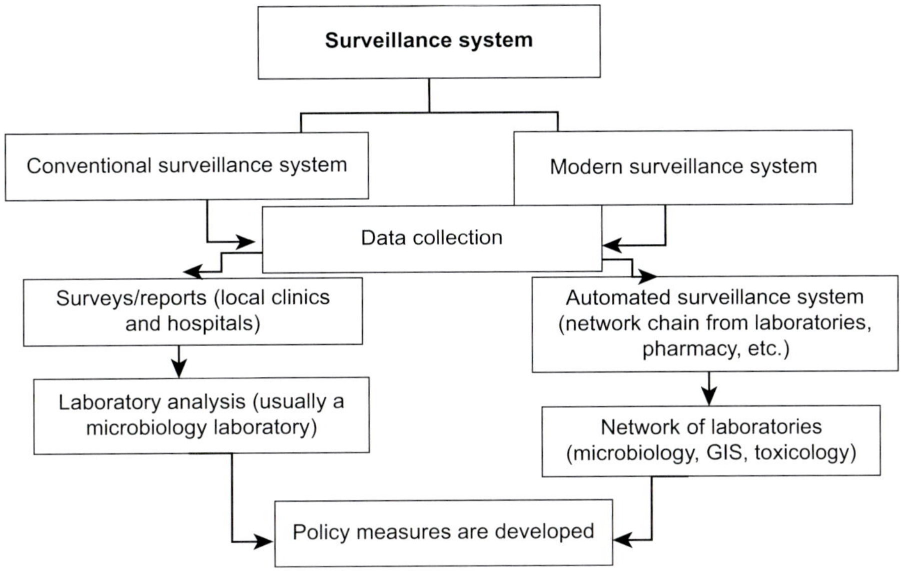

FIGURE 7.1 Steps involved in investigation of bioterrorism.

rains damaged crops in 2015. Pakistan can be beleaguered with biological weapons on the pillars of economy such as food, agriculture, human resource, and livestock (Shinwari et al., 2014).

7.3.1 Bioterrorism in health sector—case studies from Pakistan

Use of biological weapons has been widely debated over the years. World in the 21st century is more prone to bioweapons than nukes. Gleick (2006) reported cases of bioterrorism in Pakistan. Since then, numerous cases have been reported confirming the presence of evidence and traces of bioweapons, which show that specialized knowledge has been applied to make them cheap and spread to masses. Table 7.3 provides insight in the history of bioterrorism in Pakistan.

Mosquito-borne diseases have greatly affected the agricultural sector of Pakistan. Floods of 2010 and 2011 resulted in major outbreaks of water-borne diseases. According to World Health organization (WHO), 50 million dengue infections are reported each year. Dengue was first reported in 1994 in Pakistan, and till now 4.5 million cases of malaria are reported each year in Pakistan (Khan, 2016). In 2014, 1218 cases of cholera were reported

Table 7.3 List of legal parameters adopted to counter bioterrorism in Pakistan (Khalil et al., 2015).

Legal parameters	Year	Features
Biological toxic Weapons Convention (BTWC)	1972	Pakistan became a party to this convention by recognizing its obligation to prevent the use of toxic agents.
Plant Quarantine Act	1976	Pakistan is committed to regulate the entry, spread, and distribution of exotic insects and plants.
Animal Quarantine Act	1979	Pakistan is committed to regulate the entry, spread, distribution, and export of harmful animals and animal products.
Cartagena Protocol on Biosafety	1992	Being a secretary of Convention on Biological Diversity (CBD), Cartagena Protocol on Biosafety ensures safe and protected transboundary movement of biologically modified organisms.
Sanitary and Phytosanitary Agreement (committed under WHO)	1995	Pakistan is obliged to sanitary and phytosanitary agreement which coerces it not to import or export toxins which can be misused for terrorism against plants or animals.
Pakistan Environmental Protection Act (PEPA)	1997	Section 13 of PEPA prohibits the import of hazardous substances in Pakistan.
Pakistan Export Control Act	2004	Pakistan is committed to regulate, control, and prevent the spread of bioweapons. The list of such goods and technologies was prepared and revised in 2016 (confirm from the document)
National Biosafety Guidelines	2005	These guidelines were developed to regulate the possible undesirable effects of laboratory work on Genetically Modified Organisms (GMOs). These guidelines also contributed to formation of the following regulatory bodies: • Institutional Biosafety Committee (IBC) • Ministerial Biosafety Committee (MBC) • National Biosafety Committee (NBC)
International Health Regulations (committed under WHO)	2005	Pakistan is committed to regulate spread of infectious materials at inter region and intraregional level by adopting better public health response.
National Internal Security Policy (NISP)	2014	This policy addresses both internal and external threats to the country including use of biological or chemical agents for terrorism in the future.

(The Statistics portal, 2014). Each year in Pakistan, 115.2 million cases of diarrhea are reported (Sultan and Memon, 2017).

7.3.2 Pakistan's stance on countering bioterrorism

Pakistan has been a victim of bioterrorism from a very long period. Despite having no recognition in *Anti-Terrorism Act 1997* and *National Counter Terrorism Act 2013*, bioterrorism is one of the highly debated issues of Pakistan. Biosafety and bioterrorism are two parallel concepts as improvement in biosafety can help to combat bioterrorism and vice versa. Pakistan still needs to work in public health sector by developing effective response mechanisms to address bioterrorism. Up-to-date vaccine availability, clean environment, and improved health standards can be ensured by public-private partnership and joint ventures by researchers, physicians, policy makers, and common people. Various research institutes are working to protect it from this menace. A list of some initiatives from the past is given in the following section.

7.4 Bioterrorism in terms of warfare

Use of infectious materials for warfare can be traced to 600 BC to physically weaken the enemy. Use of filth, cadavers, contagion, pollutants in water reservoirs of the opposing army was a common strategy which has been followed till 21st century (Riedel, 2004). Evidence has indicated the presence of biological warfare program in Germany during World War I when Germans were accused for shipping horses and cattle inoculated with anthrax and glanders to neighboring countries (Huge-Jones, 1992). More countries invested in biological weapons during World War II. Japan conducted biological weapons research program "Unit 731" which consisted of 150 buildings, 5 satellite camps, and more than 3000 scientists (Baskin and Brewer, 1997).

State and nonstate actors have been reported to be indulged in chemical weapons programs as a prologue of terrorism in the 21st century (Rapoport, 2014). In recent era, almost 30 states have chemical weapons. Iraq used them against Iran, Libya against Chad, and Egypt used them against Yemen (Maloney, 2008). Such chemical weapons have more tactical effects than strategic ones (Bruck and Flowerree, 1991). Syrian biochemical attack is the most brutal form of bioterrorism in the recent age. Syrian government used similar weapons in 2015 and 2016 as well claiming lives of 1500 people. Thousands of people were wounded, and the blame was transferred to the rebels. In 2017, a war plane dropped a bomb in *Khan Sheikhoun*, a small area occupied by protesters. More than 70 deaths and more than 350 injuries were reported. Victims reported experiences including redness of eyes, constricted pupils, blue derma and lips, and shortness of breath. The organization for the prevention of chemical weapons collected medical samples from three victims, which indicated the presence of banned compound sarin, which is diisopropyl methylphosphonate (Syrian Civil War, 2017). This incident was a cruel decision against civilians demonstrating might is right in religious violence as well.

7.5 Case studies of bioterrorism in Pakistan

Various research studies imply that many of the diseases have been introduced as a consequence of bioterrorism in the country. Table 7.4 shows the list of distribution of deadly diseases in Pakistan (Fig. 7.2).

7.5.1 Bioterrorism in terms of health nuisance in Pakistan

Dengue fever has been identified as the worst form of epidemic across the world. Pakistan, like other developing counties, is largely affected by it due to lack of proper sanitation, unsafe access to drinking and freshwater reservoirs, larger refugees, and overcrowded cities. Its origin can be traced to import of used tires in the country. Morbidity by dengue fever has been increasing since it was first reported in 1994 (Jahan, 2011). In 2012, Government of Punjab launched anti-dengue campaign in different schools, universities, and departments. Punjab Municipal Development Fund

Table 7.4 Reported cases of bioterrorism in Pakistan.

City	Disease	Year
Lahore	Swine flu	2015
Islamabad	Crimean-Congo hemorrhagic fever	2015
Karachi	*Naegleria fowleri* (brain-eating bacteria)	2015
Noorpur Thal	Mysterious infection in 500 camels	2015
Punjab	Dengue	2011–present
Punjab	Anthrax	2004
Country wide	198 cases of polio	2011

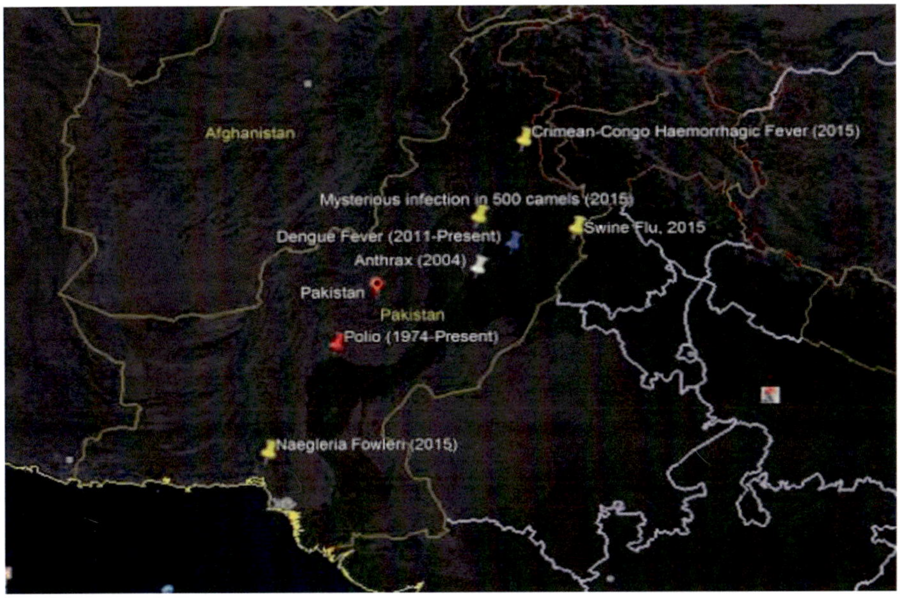

FIGURE 7.2 Highlights of the reported cases of bioterrorism in Pakistan.

Company (PMCFC) was given the responsibility to strengthen Tehsil Municipal Authorities in combating dengue. Digital maps were prepared by using geographical information system, and hotspots of dengue were identified. 520 shoulder foggers, 1600 spray pumps, and safety gears were provided to the teams who visit homes for sprays and site inspection. Various training sessions are arranged for capacity building of the staff (PMDFC, 2018). Despite numerous initiatives, dengue returns each year with larger number of cases. Lack of involvement of public, open drains, and insufficient access to sanitation can be major causes of the wide spread of this epidemic. Media has played effective role in spreading awareness about this menace; however, there are still gaps in community involvement in this sector.

7.5.2 Bird flu

Bird flu, also known as fowl plague, avian flu, or chicken Ebola, is the most fatal disease which results in 100% mortality rate in domesticated birds (Dhama et al., 2013). H5N1 or etiological agent avian influenza virus (AIV) has global impacts and spreads beyond geographical boundaries (Dhama et al., 2012). Therefore, disease outbreak should be reported immediately so that trained veterinarians and professionals can deal with biohazard of fowl birds. The symptoms of bird flu in fowls include edema in wattles, diarrhea, soft-shelled or misshaped eggs, lack of coordination in birds, swelling of birds, lack of appetite or energy, ruffled feathers, and nasal discharge (Bentoli, 2018). In 2003, human to human transmission of avian influenza was observed before which it was considered to affect birds only (Peiris et al., 2004).

The poultry sector of Pakistan also witnessed episodes of H5N1, which resulted in great challenges to this industry due to highly pathogenic Asian influenza (HPAI) during 2003–06 (Ghafoor et al., 2010). Avian influenza was first recorded in 2006 in Pakistan in Charsadda, Abbottabad, Pindi Bhatian, Attock, Sehala, and Tarlai areas of Islamabad. Nearly 100,000 birds were culled, and Rs. 100 million was spent as compensation. The surveillance system established nationwide helped stakeholders to prevent disasters from this disease (Hussain et al., 2008). CDC advises the travelers and people who deal it to take prevention measures (Khan et al., 2008). Pakistan gained the bird flu free status in 2007 (Poultry News, 2011).

7.6 Conclusion

Bioterrorism is real, evident, and persistent in the modern era. Unfortunately, Pakistan faces bioterrorism in all categories. Our natural resources are being polluted at irreversible rate. Many indigenous species are at the verge of extinction. Evidence shows that state and nonstate actors are contributing to biosafety and security issues. Therefore, it is important to revise and review the legal structure pertaining bioterrorism. Government must ensure the construction of strong institutions which can carefully implement those legislations. The source of category A and B is localized; therefore strengthening of provincial Environmental Protection Agencies (EPA), implementation of self-monitoring programs by the industries, and legislations such as "polluter pays" principle can help to protect our natural resources. Research is the backbone of policies. Therefore, private sector should be encouraged to invest in research in this field.

References

Barras, V., Greub, G., 2014. History of biological warfare and bioterrorism. Clinical Microbiology and Infections 20 (6), 497–502.

Baskin, S.I., Brewer, T.G., 1997. In: Sidell, F., Takafuji, E.T., Franz, D.R. (Eds.), Medical Aspects of Chemical and Biological Warfare, vol. 10. TMM Publication, Washington, DC, p. 271.

Bentoli, 2018. How to Identify Bird Flu Symptoms in Chicken. https://bentoli.com/bird-flu-symptoms-in-chickens/.

Bravata, D.M., McDonald, K.M., Smith, W.M., Rydzak, C., Szeto, H., Buckeridge, D.L., Haberland, C., Owens, D.K., 2004. Systematic review: surveillance systems for early detection of bioterrorism-related diseases. Annals of Internal Medicine 140 (11), 910−922.

Brooks, P., 2007. Bird Flu (Avian Flu). http://www.netdoctor.co.uk/healthy-living/a5566/bird-flu-avian-flu/.

Burck, G.M., Flowerree, C.C., 1991. International Handbook on Chemical Weapons Proliferation, p. 105. New York.

Çankaya, S., Kibaroğlu, M., 2010. Bioterrorism. IOS Press, Amsterdam.

CDC, 2018. Data and Statistics about Tuberculosis Around the World. https://cdc.gov/tb/statistics/default.htm/.

Clercq, E., Kern, E., 2003. Handbook of Viral Bioterrorism and Biodefence. Elsevier, Amsterdam..

Dhama, K., Chakraborty, S., Tiwari, R., Kumar, A., Rahal, A., Latheef, S.K., Latheef, Wani, M.,Y., Kapoor, S., 2013. Avian/bird flu virus: poultry pathogen having zoonotic and pandemic threats—a review. Journal of Medical Science 13 (5), 301−315.

Dhama, K., Tiwari, R., Singh, S.D., 2012. Biosecurity measures at poultry farms and thumb rules to avoid developing a serious zoonotic illness from birds. Poultry Punch 28 (3), 30−51.

Ghafoor, A., Badar, H., Hussain, M., Tariq, N., 2010. An empirical estimation of the factors affecting demand and supply of poultry meat. Pakistan Veterinary Journal 30 (3), 172−174.

Gholipour, B., 2013. 2009 Swine-Flu Death Toll 10 Times Higher than Thought. https://www.livescience.com/41539-2009-swine-flu-death-toll-higher.html.

Gleick, P.H., 2006. Water and terrorism. Water Policy 8 (6), 481−503.

Henderson, D.A., 1998. Bioterrorism as a public health threat. Emerging Infectious Diseases 4 (3), 488−495.

Hugh-Jones, M., 1992. Wickham Steed and German biological warfare research. Intelligence and National Security 7 (4), 379−402.

Hussain, M., Irshad, H., Khan, M.Q., 2008. Laboratory diagnosis of transboundary animal diseases in Pakistan. Transboundary and Emerging Diseases 55 (5-6), 190−195.

Hutwagner, L., Thompson, W., Seeman, G.M., Treadwell, T., 2003. The bioterrorism preparedness and response early aberration reporting system (EARS). Journal of Urban Health 80 (1), i89−i96.

Jahan, F., 2011. Dengue fever (DF) in Pakistan. Asia Pacific Family Medicine 10 (1), 1.

Jaspal, Z.N., 2010. Biosecurity and Pakistan: A Critical Appraisal.

Khalil, A., Tanveer, F., Shinwari, Z.K., 2015. Pakistan's bio-preparedness with regard to biosecurity, biodefense strategies and policy measures. Journal of Bioterrorism & Biodefense 6 (132), 2.

Khan, S., 2016. Malaria in Pakistan. Pakistan Observer. https://pakobserver.net/malaria-in-pakistan/.

Khan, S.A., Hashmi, I., Ranjha, A.N., Yosufzai, M.K., Hashmi, K., Anjum, Q., 2008. Students' corner-awareness of bird flu amongst young college students. The Journal of the Pakistan Medical Association 58 (8), 466.

Khardori, N., 2006. Bioterrorism and bioterrorism preparedness: historical perspective and overview. Infectious Disease Clinics 20 (2), 179–211.

Lim, D.V., Simpson, J.M., Kearns, E.A., Kramer, M.F., 2005. Current and developing technologies for monitoring agents of bioterrorism and biowarfare. Clinical Microbiology Reviews 18 (4), 583–607.

Maloney, S., 2008. Iran's Long Reach: Iran as a Pivotal State in the Muslim World. US Institute of Peace Press.

Mikes, J., 2009. Bioterrorism in Asia. The Diplomat. https://thediplomat.com/2009/05/bioterrorism-in-asia/.

Nazar, S., 2016. Pakistan's Big Threat Isn't Terrorism-It's Climate Change. The South Asian Channel. http://foreignpolicy.com/2016/03/04/pakistans-big-threat-isnt-terrorism-its-climate-change/.

Peiris, J.S.M., Yu, W.C., Leung, C.W., Cheung, C.Y., Ng, W.F., Nicholls, J.A., Path, T.,N., Chan, K., Lim, W., Yuen, K., Guan, Y., 2004. Re-emergence of fatal human influenza A subtype H5N1 disease. The Lancet 363 (9409), 617–619.

PMDFC, 2018. Anti-Dengue Campaign. http://pmdfc.punjab.gov.pk/anti_dengue_campaign/.

Rapoport, D.C., 2014. Terrorism and weapons of the apocalypse. In: Twenty-first Century Weapons Proliferation. Routledge, pp. 32–50.

Riedel, S., 2004. Biological warfare and bioterrorism: a historical review. In: Baylor University Medical Center Proceedings, vol. 17, No. 4. Taylor & Francis, pp. 400–406.

Ryan, J., Glarum, J., 2008. Biosecurity and Bioterrorism. Butterworth-Heinemann.

Shinwari, Z.K., Khalil, A.T., Nasim, A., 2014. Natural or deliberate outbreak in Pakistan: how to prevent or detect and trace its origin: biosecurity, surveillance, forensics. Archivum Immunologiae et Therapiae Experimentalis 62 (4), 263–275.

Sultan, M.A., Memon, I., 2017. 5th Initiative against Diarrheal and Enteric Diseases in Asia. http://www.cholera-network.org/wp-content/uploads/2017/05/ashraf-sultan.pdf/.

Syrian Civil War, 2017. Syrian Chemical Attack: What We Know. http://www.bbc.com/news/world-middle-east-39500947/.

The Poultry Site, 2011. Vigilance & Prevention Keep Pakistan Bird Flu- Free. http://www.thepoultrysite.com/poultrynews/22942/vigilance-prevention-keep-pakistan-bird-flundashfree/.

The Statistical Portal, 2014. Number of Reported Cases of Cholera in Pakistan. https://www.statista.com/statistics/701551/pakistan-number-of-reported-cases-of-cholera/.

Index

'*Note:* Page numbers followed by "f" indicate figures and "t" indicate tables.'

D

E

F

G

H

I

Kolibri 360 GmbH
Ferdinand-Jühlke-Straße 7
99095 Erfurt, Deutschland
produktsicherheit@kolibri360.de